Learning Materials in Biosciences

Learning Materials in Biosciences textbooks compactly and concisely discuss a specific biological, biomedical, biochemical, bioengineering or cell biologic topic. The textbooks in this series are based on lectures for upper-level undergraduates, master's and graduate students, presented and written by authoritative figures in the field at leading universities around the globe.

The titles are organized to guide the reader to a deeper understanding of the concepts covered.

Each textbook provides readers with fundamental insights into the subject and prepares them to independently pursue further thinking and research on the topic. Colored figures, step-by-step protocols and take-home messages offer an accessible approach to learning and understanding.

In addition to being designed to benefit students, Learning Materials textbooks represent a valuable tool for lecturers and teachers, helping them to prepare their own respective coursework.

More information about this series at http://www.springer.com/series/15430

Melanie Kappelmann-Fenzl

Editor

Next Generation Sequencing and Data Analysis

 Springer

Editor
Melanie Kappelmann-Fenzl
Faculty of Applied Informatics
Deggendorf Institute of Technology
Deggendorf, Germany

ISSN 2509-6125   ISSN 2509-6133   (electronic)
Learning Materials in Biosciences
ISBN 978-3-030-62489-7   ISBN 978-3-030-62490-3   (eBook)
https://doi.org/10.1007/978-3-030-62490-3

This Springer imprint is published by the registered company Springer Nature Switzerland AG.
The registered company address is: Gewerbestrasse 11, 6330 Cham, Switzerland

Preface

The intention to write this textbook came from the fact that when I started to deal with NGS data analysis, I felt like I was abandoned in a jungle. A jungle of databases, scripts in different programming languages, packages, software tools, repositories, etc. What you need from all these components for what purpose of NGS analysis, I had to painstakingly learn from numerous tutorials, publications (and there are plenty of them), books and by attending courses. I also had to find a lot of information on specific problem solutions in bioinformatics forums such as *Biostars* (https://www.biostars.org/). A basic initial structure regarding the "right" approach to get solid results from the NGS data on a defined problem would have been more than desirable. This textbook is intended to facilitate these circumstances and introduce NGS technology, its application and the analysis of the data obtained. However, it should be kept in mind that this textbook does not cover all possibilities of NGS data analysis, but rather provides a theoretical understanding, based on a few practical examples, to give a basic orientation in dealing with biological sequences and their computer-based analysis.

Computational biology has developed rapidly over the last decades and continues to do so. Bioinformatics not only deals with NGS data, but also with molecular structures, enzymatic activity, medical and pharmacological statistics, to name just a few topics. Nevertheless, the analysis of biological sequences has become a very important part in bioinformatics, both in the natural sciences and in medicine. In order to enable you to handle NGS data professionally and to be able to answer specific research questions about your sequencing data without having to give your data into other hands and to invest a lot of money for it, this textbook is meant to be a practical guide. In addition, the experimental procedure (library preparation) required prior to the actual sequencing as well as the most common sequencing technologies currently available on the market is also illustrated.

Protocols should be viewed as guidelines, not as rules that guarantee success. It is still necessary to think by yourself—NGS data analysts need to recognize the complexity of living organisms, respond dynamically to variations, and understand when methods and protocols are not suited to a data set. Therefore, a detailed documentation of all analysis steps carried out with a note of the reason why the respective step was taken is absolutely essential. Moreover, you should really be familiar with the theoretical background of the performed working steps, the programs will not tell you that your settings will lead to incorrect results. Thus, being a data analyst is much more than just appending commands in a terminal.

Deggendorf, Germany   Melanie Kappelmann-Fenzl

Legend: How to Read this Textbook

commands that are executed in the terminal

commands that are executed in R

How to Read this Book

Additive and supporting scripts regarding NGS data analysis can be viewed at https://github.com/mkappelmann or https://github.com/grimmlab/BookChapter-RNA-Seq-Analyses. As the field of bioinformatic data analysis is rapidly evolving, you will also find the latest changes to the scripts described in the text book.

Contents

Contributors

Richa Barthi University of Applied Sciences Weihenstephan-Triesdorf, TUM Campus Straubing for Biotechnology and Sustainability, Straubing, Germany

Patricia Basurto-Lozada Laboratorio Internacional de Investigación sobre el Genoma Humano, Universidad Nacional Autónoma de México, Santiago de Querétaro, México

Anja Bosserhoff Institute of Biochemistry (Emil-Fischer Center), Friedrich-Alexander University Erlangen-Nürnberg, Erlangen, Germany

Carolina Castañeda-García Laboratorio Internacional de Investigación sobre el Genoma Humano, Universidad Nacional Autónoma de México, Santiago de Querétaro, México

Marius Eisele Faculty of Applied Informatics, Deggendorf Institute of Technology, Deggendorf, Germany

Dominik Grimm University of Applied Sciences Weihenstephan-Triesdorf, TUM Campus Straubing for Biotechnology and Sustainability, Straubing, Germany

Melanie Kappelmann-Fenzl Faculty of Applied Informatics, Deggendorf Institute of Technology, Deggendorf, Germany

Christian Molina-Aguilar Laboratorio Internacional de Investigación sobre el Genoma Humano, Universidad Nacional Autónoma de México, Santiago de Querétaro, México

Rebeca Olvera-León Laboratorio Internacional de Investigación sobre el Genoma Humano, Universidad Nacional Autónoma de México, Santiago de Querétaro, México

Raúl Ossio Laboratorio Internacional de Investigación sobre el Genoma Humano, Universidad Nacional Autónoma de México, Santiago de Querétaro, México

Carla Daniela Robles-Espinoza Laboratorio Internacional de Investigación sobre el Genoma Humano, Universidad Nacional Autónoma de México, Santiago de Querétaro, México

Experimental Cancer Genetics, Wellcome Sanger Institute, Cambridge, UK

Martha Estefanía Vázquez-Cruz Laboratorio Internacional de Investigación sobre el Genoma Humano, Universidad Nacional Autónoma de México, Santiago de Querétaro, México

Next Generation Sequencing (NGS): What Can Be Sequenced?

1

Anja Bosserhoff and Melanie Kappelmann-Fenzl

Contents

What You Will Learn in This Chapter

The content of the following chapter briefly describes the basics of biological genetic information and their function in living organisms. You will gain principle knowledge on DNA and RNA and understand the differences of analyzing these by sequencing templates. Next generation sequencing technologies can be used to

(continued)

A. Bosserhoff
Institute of Biochemistry (Emil-Fischer Center), Friedrich–Alexander University Erlangen–Nürnberg, Erlangen, Germany

M. Kappelmann-Fenzl (✉)
Institute of Biochemistry (Emil-Fischer Center), Friedrich–Alexander University Erlangen–Nürnberg, Erlangen, Germany

Deggendorf Institute of Technology, Deggendorf, Germany
e-mail: melanie.kappelmann-fenzl@th-deg.de

© Springer Nature Switzerland AG 2021
M. Kappelmann-Fenzl (ed.), *Next Generation Sequencing and Data Analysis*, Learning Materials in Biosciences, https://doi.org/10.1007/978-3-030-62490-3_1

sequence DNA or RNA, respectively, answering the research or clinical question you have. Different sequencing approaches thus lead to different readouts.

1.1   Introduction

Why do some people become seriously ill, others remain healthy throughout their lives? Why does a disease progress so differently in different people? Why does a drug work optimally in one patient and not in others? The answer to these questions lies in the complex individuality of each person and a medicine that does justice to it—personalized medicine. To examine the healthy as well as the sick person in the finest detail and to calculate the results with a lot of computer capacity to a meaningful image helps to understand and to treat more precisely also particularly complex illnesses, such as psychiatric disorders, cardiovascular diseases, inflammatory diseases, or also cancer. Our genetic pattern, but also differences in our diet, environment, or lifestyle have an effect on our state of health. How do the individual factors contribute to a disease and how can they be influenced? It is obvious that in the case of complex diseases one cannot just consider individual factors. Only investigations of the exact interplay and the chronological sequence enable a deeper understanding of the health and illness of the human body. Today, the power of computers with enormous computing capacity is used to determine complex relationships of different influences from detailed measurements on humans, to create mechanistic models being then tested in the laboratory for their accuracy. In this process, different levels of data are examined, ranging from single clinical observations to complicated molecular data sets. The approach known as "systems medicine" uses the quantities of data by relating them intelligently, designing predictive models and thus helping to develop innovative therapeutic and preventive procedures [1–5].

1.2   Biological Sequences

In general, you can say that living organisms consist of cells, which share common features but also differ in function and morphology. Anyway, almost every eukaryotic cell type carries a nucleus harboring our genetic make-up—the DNA. The DNA stores all the information, which is essential for producing, e.g., a human being, similar to an instruction manual. To make the information written in the DNA usable, parts of the DNA are transcribed into another kind of biological information—the RNA. The parts of the DNA, which are transcribed into RNA, are called coding regions or genes. Thus, the RNA transports the information of the DNA out of the nucleus (this RNA is, therefore, called "messenger RNA" (mRNA)). A significant portion of the RNA is then used as information being translated into another biological molecule—a protein. Proteins are composed of amino acids and fulfill almost all structural, regulatory, and signal transducing functions in the body. RNAs and thus proteins are, therefore, essential mediators of what is written in our instruction manual DNA. Which parts of the DNA are transcribed into RNA and finally translated into proteins vary between the different cell types, depending on the cell's differentiation status. Thus, for example, a brain cell (neuron) shows a different gene expression pattern than a liver cell (hepatocyte) (the gene expression pattern is thereby also the reason to be a neuron and hepatocyte, respectively). Which genes are expressed in cells is highly regulated by multiple complex molecular mechanisms affecting either the DNA, RNA, or protein level [6, 7]. The different possible sequencing structures, their origin, and NGS applications are depicted in Fig. 1.4.

1.2.1   DNA

DNA stands for *Deoxyribonucleic acid* and is a macromolecule (composed of smaller molecules). The backbone of the DNA consists of alternating molecules: deoxyribose and phosphate, complemented by the important nucleobases A (Adenine), C (Cytosine), G (Guanine), and T (Thymine) bound to the deoxyribose. Thus, the genetic code is written by those four letters. The DNA molecule is double-stranded, orientated antiparallel, and 3-dimensionally organized as a helix.

The nucleotides are held together by intermolecular hydrogen bonds between the two strands, whereby A only pairs with T and G only pairs with C (complementary base pairing). Consequently, if you know the sequence of one strand, you automatically know the sequence of the other one (Fig. 1.1). The directionality of the DNA is determined by the polarity of the molecules and is given in 5' to 3' direction (Fig. 1.1). Thus, a DNA sequence TGCCA may need to be considered:

- in reverse, ACCGT
- as a complement, ACGGT
- as a reverse complement, TGGCA

The two strands are labeled arbitrary:

- positive/negative; +/−
- positive/negative

5′	A	C	T	G	A	C	C	G	A	A	3′
	|	|	|	|	|	|	|	|	|	|	
3′	T	G	A	C	T	G	G	C	T	T	5′

Fig. 1.1   Schematic illustration of DNA

- forward/reverse
- top/bottom
- Watson/Crick
- leading/lagging, etc.

It should be noted that the different possible labels of the DNA strands should not be confused with the directionality sense and antisense. The antisense strand *is* transcribed into RNA not the sense strand, resulting in an RNA having the sequence of the sense strand and being the reverse complement of the antisense strand (Fig. 1.2). However, the sequence may come from the forward or reverse strand.

The main purpose of the DNA or genome is to make the functioning of a living organism possible. Therefore, the DNA carries genes (coding regions) containing the instruction for making proteins or other molecules like non-coding, but functional or regulatory, RNAs. But protein production is a little more complex: the coding regions on the DNA are composed of a so-called intron–exon structure, whereby only the exons are essential for the coding transcript and for possible protein production. Hence, the introns have to be removed (splicing) in the transcribed RNA. This fact enables to produce a variety of proteins out of one gene by a mechanism called alternative splicing (Fig. 1.2). The "official" definition of a gene is [8]:

A region (or regions) that includes all of the sequence elements necessary to encode a functional transcript. A gene may include regulatory regions, transcribed regions and other functional sequence regions.

Thus, talking about genes means, in most cases, talking about genomics. The human genome consists of three billion base pairs (3Gb), in comparison the genome of a fruit fly

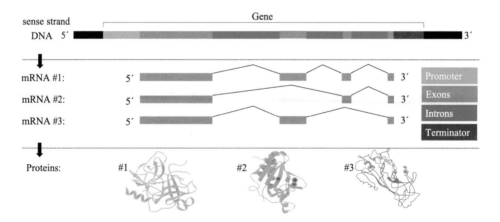

Fig. 1.2 Simplified representation of gene transcription resulting in three different mRNA transcripts via alternative splicing, which are then translated into three different proteins

consists of 120 million base pairs (120 Mb). Due to the huge amount of data, it is common to refer to genome sizes in terms of kilo-bases (thousands), mega-bases (millions), and giga-bases (billions). But there is absolutely no relation between genome size and complexity or intelligence of an organism.

1.2.1.1 RNA

As already mentioned above, the DNA is written in a four-letter code (T/A/C/G) partially containing genes. In eukaryotic cell systems, the DNA never leaves the nucleus, thus a structure called RNA is created, a transcript of the DNA, carrying the genetic information out of the nucleus, e.g., to guarantee translation into functional proteins. The RNA code is similar to the genetic one; however, RNA is single stranded, the nucleotide T is replaced by a U (Uracil) and after transcription from DNA to RNA, we do not longer talk about genomics, but transcriptomics. Several different kinds of RNA exist (Table 1.1). One RNA, the kind of RNA molecule harboring the coding information of the DNA (genes) to produce proteins (mRNA), is processed in eukaryotes: After leaving the nucleus the intronic structures are removed and the molecules are ready for translation. Due to the fact that these RNA molecules carry genetic information out of the nucleus they are called messenger RNAs (Fig. 1.2). Splicing events (removal of introns) can be attributed to the splice signals GT and AG (Fig. 1.3), but the presence of these dinucleotides does not mean that splicing is mandatory (alternative splicing; see Fig. 1.2).

Table 1.1 Types of RNA molecules

Abbreviation	Name	Function
mRNA	Messenger RNA	coding for protein; ~5% of total RNA
rRNA	Ribosomal RNA	Central structure of ribosomes; catalyzes the formation of peptide bonds during protein synthesis; ~80% of total RNA
tRNA	Transfer RNA	Adapters between mRNA code and amino acids during proteins synthesis inside of ribosomes; ~15% of total RNA
snRNA	Small nuclear RNA	Essential for RNA splicing
snoRNA	Small nucleolar RNA	Guides for RNA modification and processing
miRNA	Micro RNA	Blocking the translation of distinct mRNAs
siRNA	Small interfering RNA	Mediates degradation of distinct mRNAs and closing of gene loci leading to decreased gene expression

mRNA: 5′

Fig. 1.3 Splice signals usually occur as the first and last dinucleotides of an intron

1.2.2 Protein

Once the genetic code is transcribed into an RNA molecule and the intronic regions are removed, the mRNA can be translated into a functional protein. The molecular process from a mRNA to a protein is called translation. Proteins consist of amino acids and are, as well as DNA and some RNAs, formed 3-dimensionally. There are generally 20 different amino acids, which can be used to build proteins. An amino acid (AA) chain with less than 40 AAs is called a polypeptide. There are 64 possible permutations of three-letter sequences that can be made from the four nucleotides. Sixty-one codons represent amino acids, and three are stop signals. Although each codon is specific for only one amino acid (or one stop signal), a single amino acid may be coded for by more than one codon. This characteristic of the genetic code is described as degenerate or redundant. The genetic codons are illustrated in Table 1.2, representing all nucleotide triplets and their associated amino acid(s), START or STOP signals, respectively. Just like DNA and RNA, a protein can also be described by its sequence, however, the 3-dimensional structure based on the biochemical properties of the amino acids and the milieu, in which proteins are folding, are much more essential for those building blocks of life.

To keep complex things simple, the way our genes (DNA) are converted to a transportable messenger system (mRNA) to a functional protein is illustrated in Fig. 1.4 [9].

The above-mentioned molecular structures can be analyzed in detail by the Next Generation Sequencing (NGS) technology, enabling a genome wide insight into the organization and functionality of the genome and all other molecules resulting therefrom [10]. These will be explained in the following chapters.

1.2.3 Other Important Features of the Genome

Before we can go into detail in terms of sequencing technologies and application, we have to mention some other important features of the genome and its associated molecules. As you already know, the human genome consists of ~3 billion base pairs and is harbored in almost every single cell within the human organism (~10^{14} cells). Actually, that is quite a lot. However, only roughly 1.5% of the whole genome is coding for proteins. What about the rest? Ninety-eight percent of the human genome useless? Obviously not. The ENCODE (Encyclopedia of DNA Elements) project has found that 78% of non-coding DNA serve a defined purpose [11]. Well, think about all the different tasks of all the different cell types making up a human being. Not every single cell has to be able to do everything—they are

Table 1.2   The genetic code. The biochemical properties of the amino acids are colored: nonpolar (red), polar (green), basic (violet), acidic (blue)

UUU	Phenylalanine (Phe/F)	UCU	Serine (Ser/S)	UAU	Tyrosine (Tyr/Y)	UGU	Cysteine (Cys/C)
UUC		UCC		UAC		UGC	
UUA	Leucine (Leu/L)	UCA		UAA	Stop	UGA	Stop
UUG		UCG		UAG		UGG	Tryptophan (Trp/W)
CUU		CCU	Proline (Pro/P)	CAU	Histidine (His/H)	CGU	Arginine (Arg/R)
CUC		CCC		CAC		CGC	
CUA		CCA		CAA	Glutamine (Gln/Q)	CGA	
CUG		CCG		CAG		CGG	
AUU	Isoleucine (Ile/I)	ACU	Threonine (Thr/T)	AAU	Asparagine (Asn/N)	AGU	Serine (Ser/S)
AUC		ACC		AAC		AGC	
AUA		ACA		AAA	Lysine (Lys/K)	AGA	Arginine (Arg/R)
AUG	Methionine (Start, Met/M)	ACG		AAG		AGG	
GUU	Valine (Val/V)	GCU	Alanine (Ala/A)	GAU	Aspartate (Asp/D)	GGU	Glycine (Gly/G)
GUC		GCC		GAC		GGC	
GUA		GCA		GAA	Glutamate (Glu/E)	GGA	
GUG		GCG		GAG		GGG	

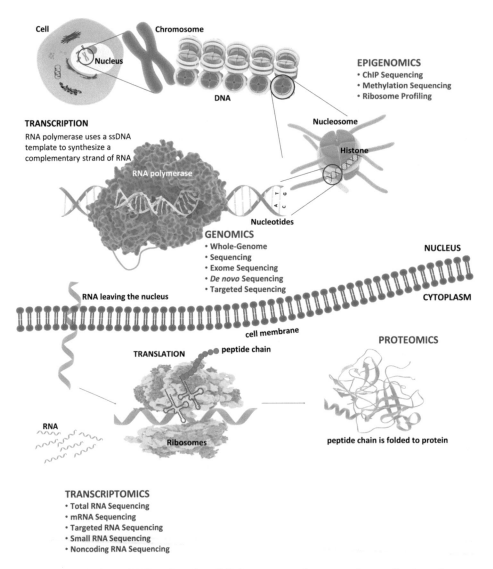

Fig. 1.4 From DNA to RNA to Proteins. Cellular structures for sequencing applications. (source: © Melanie Kappelmann-Fenzl)

specialized. Thus, different cell types need different genes in the whole genome to fulfill their job within the body or to build certain structures. Consequently, the human genome has a kind of switch on/off mechanism. Genes, which are needed are transcribed and those, which are not needed are not. To regulate this, on the one hand, regulatory sequences can be found in the genome, like silencer, enhancer, promoter regions. On the other hand, epigenetic mechanism can influence gene expression and thus protein production.

In fact, there are various regulatory mechanisms, which precisely control gene expression. In the following we just want to mention the most common once:

- *Promoter*: Is a genomic region, where the transcription machinery can bind to transcribe the following gene into an RNA molecule. Often, promoter regions can be associated with a high CG content.
- *Terminator*: Is a genomic region, where the transcription process, and thus the gene transcript, ends.
- *Silencer*: A genomic pattern that decreases the frequency of transcription and thus the expression of the regulated gene.
- *Enhancer*: A genomic pattern that increases the frequency of transcription and thus the expression of the regulated gene.
  Silencer as well as enhancer DNA can be far from the gene in a linear way; however, it is spatially close to the promoter and gene. This is managed by folding of the DNA.
- *CpG islands*: Genomic regions, which are enriched by Cs followed by a G (CpG, p stands for the phosphate in the DNA backbone). Cytosines in CpG islands can be methylated, which is an epigenetic regulatory mechanism of gene expression. In most cases DNA methylation is associated to the inactivation of the corresponding genes.
- *Epigenetics*: Includes CpG island DNA methylation as mentioned above, but also histone modifications influencing transcriptional activity.
- *UTRs* (untranslated regions): The 5'UTR is located before the start codon and the 3'-UTR is the region between the stop codon and the poly-A-tail of the mRNA.

In addition to the regulatory sequences, there are various other sequence components making up the human genome [12]. The main components are illustrated in Fig. 1.5.

Basically, NGS applications are performed on DNA and RNA molecules. Sequencing these molecules enables us to identify DNA and RNA sequences but also to define, e.g., DNA–protein interactions or epigenetic DNA modifications. Hence, NGS output data give a great insight into structural and functional characteristics of cells and tissues. Each NGS application can give a different result, depending on the specific research question. Rapid DNA and RNA sequencing is now mainstream and will continue to have an increasing impact on biology and medicine [13].

Common NGS applications are:

- *Expression analysis*
  The RNA-Seq application enables you to investigate expression variances of RNA structures of, e.g., different tissues. Moreover, RNA-Seq reads can be used to analyze differential exon usage, gene fusions or variants like SNPs, indels, mutations, etc. [14].
- *DNA–protein interactions*
  The ChIP-Seq (*ch*romatin-*i*mmuno*p*recipitation) application focuses on investigating regulatory sequences of the DNA, like transcription factor binding sites or histone modifications (e.g., acetylation/methylation), which lead to differences in gene

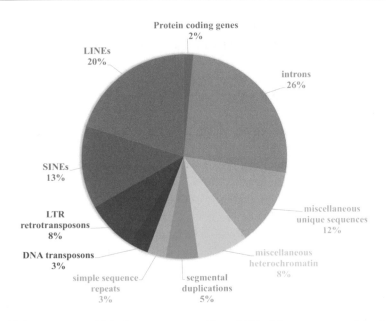

Fig. 1.5 Main components of the human genome. About 1.5% of the genome consists of ~20.000 protein-coding sequences interspersed by the non-coding introns (~26%). The largest fraction (40–50%) consists of the transposable elements, including long interspersed nuclear elements (LINEs), and short interspersed nuclear elements (SINEs). Most transposable elements are genomic remains that are currently defunct (modified according to Lander et al. 2001)

expression profiles. These regulatory sequences on the DNA are often localized in the so-called promoter, silencer, or enhancer regions and effect the transcription of nearby positioned genes [15].

- *DNA methylation (methyl-Seq)*

  Similar to the ChIP-Seq application, the power of NGS gave a boost to the study of DNA methylation in CpG islands. DNA methylation is an epigenetic modification that plays an essential role in regulating gene expression and allelic use and, consequently, influences a wide variety of biological processes and diseases [16].

- *Whole genome sequencing*

  Whole genome sequencing (WGS) is a complete readout of all the information that makes up all of your DNA. WGS gives you a deep insight into chromosomal alterations, indels, mutations, etc. [17].

- *Whole-exome sequencing*

  The exome comprises just over 1% of the whole genome and is providing sequence information for protein-coding regions. This application is widely used to identify disease-causing mutations [18].

- *Targeted sequencing*

  Targeted sequencing allows to focus on specific genes or regions of interest and thus enables sequencing at much higher coverage levels. There are a variety of targeted sequencing panels available, however, it is also possible to generate custom probe sets [19]. Targeted sequencing is mainly used in molecular diagnostics utilizing the so-called *tumor panels* (specific to the respective tumor entity).
- *De novo sequencing*

  This method refers to sequence a novel genome, if there is no reference genome available for alignment [20, 21].

This list represents only a small subset of the possible NGS applications. A detailed exposition of the DNA, RNA, and epigenetic sequencing methods can be found under the following links:

- DNA Sequencing Methods Collection:
    - https://emea.illumina.com/content/dam/illumina-marketing/documents/products/research_reviews/dna-sequencing-methods-review-web.pdf
    - https://www.thermofisher.com/de/de/home/life-science/sequencing/dna-sequencing/whole-genome-sequencing/whole-genome-sequencing-ion-torrent-next-generation-sequencing.html
    - https://www.pacb.com/applications/whole-genome-sequencing/
    - https://nanoporetech.com/applications/whole-genome-sequencing
- RNA Sequencing Methods Collection:
    - https://emea.illumina.com/content/dam/illumina-marketing/documents/products/research_reviews/rna-sequencing-methods-review-web.pdf
    - https://www.thermofisher.com/de/de/home/life-science/sequencing/rna-sequencing/transcriptome-sequencing/transcriptome-sequencing-ion-torrent-next-generation-sequencing.html
    - https://www.pacb.com/applications/rna-sequencing/
    - https://nanoporetech.com/applications/rna-sequencing
- Methylation Sequencing:
    - https://emea.illumina.com/techniques/sequencing/methylation-sequencing.html
    - https://www.thermofisher.com/de/de/home/life-science/sequencing/epigenetic-sequencing/methylation-analysis.html
    - https://www.pacb.com/applications/epigenetics/

Take Home Message
- The flow of genetic information within a biological system can be shortly described by "DNA makes RNA and RNA makes Protein."
- The duplication of DNA during the cell cycle and thus cell division is known as DNA-replication.
- The molecular process, in which a particular segment of DNA is converted into RNA, is called transcription.
- The molecular process, in which the genetic code in mRNA is read to make a protein, is called translation.
- Sequence information of the different "Omics" levels provide deep insights into molecular mechanisms of living organisms.
- Each NGS application results in a different readout, depending on the specific research question.
- The commonly used NGS applications are RNA-Seq, ChIP-Seq, methyl-Seq, WGS, WES, targeted sequencing, and *de novo* sequencing.

Further Reading

- Clancy S, Brown W. Translation: DNA to mRNA to Protein. Nature Education. 2008;1 (1):101.
- Brown TA. Genomes. 2nd ed. Oxford: Wiley-Liss; 2002. Available from: https://www.ncbi.nlm.nih.gov/books/NBK21128/.
- Alberts B, Johnson A, Lewis J, Raff M, Roberts K, Walter P. *Molecular Biology of the Cell.* 7th ed. New York: Garland Science; 2002.

Review Questions

Review Question 1

 Original DNA Sequence: 5'-ATGTGGAACCGCTGCTGA-3'

 Mutated DNA Sequence: 5'-ATGCTGGAACCGCTGCTGA-3'

a) What's the mRNA sequence?
b) What will be the amino acid sequence?
c) Will there likely be effects?
d) What kind of mutation is this?

Review Question 2

All cells in a multicellular organism have normally developed from a single cell and share the same genome, but can nevertheless be wildly different in their shape and function. What in the eukaryotic genome is responsible for this cell-type diversity?

Review Question 3

Indicate if each of the following descriptions matches RNA (R) or DNA (D). Your answer would be a five-letter string composed of letters R and D only.

- It is mainly found as a long, double-stranded molecule.
- It contains the sugar ribose.
- It normally contains the bases thymine, cytosine, adenine, and guanine.
- It can normally adopt distinctive folded shapes.
- It can be used as the template for protein synthesis.

Review Question 4

Imagine a segment of DNA (within a gene) encoding a certain amount of information in its nucleotide sequence. When this segment is fully transcribed into mRNA and then translated into protein, in general, . . .

A. the protein sequence would carry more information compared to the DNA and mRNA sequences, because its alphabet has 20 letters.
B. the protein sequence would carry less information compared to the DNA and mRNA sequences, because several codons can correspond to one amino acid.
C. the amount of information in the mRNA sequence is lower, because the mRNA has been transcribed using only one of the DNA strands as the template.
D. the amount of information in the mRNA sequence is higher, because several mRNA molecules can be transcribed from one DNA molecule.

Answers to Review Questions

Answers to Question 1

a) What is the mRNA sequence?

5'-AUGUGGAACCGCUGCUGA-3'
5'-AUGCUGGAACCGCUGCUGA-3'

b) What will be the amino acid sequence?

NH_3^+ - Met- Trp- Asn- Arg- Cys- Stop
NH_3^+ - Met- Leu- Glu- Pro- Leu- Leu- COO^-

c) Will there likely be effects?

Yes

d) What kind of mutation is this?

   Insertion, frame-shift

   Answer to Question 2
   Genes encoding regulatory proteins, regulatory sequences that control the expression of genes, genes coding for molecules involved in receiving cellular signals and genes that code for molecules involved in sending cellular signals to other cells.

   Answer to Question 3
   DRDRR.

   Answer to Question 4
   B.

Acknowledgements We are grateful to Dr. Ines Böhme (Institute of Biochemistry (Emil-Fischer Center), Friedrich–Alexander (University Erlangen–Nürnberg, Erlangen, Germany) for critically reading this text and correcting our mistakes.

References

1. Wakai T, Prasoon P, Hirose Y, Shimada Y, Ichikawa H, Nagahashi M. Next-generation sequencing-based clinical sequencing: toward precision medicine in solid tumors. Int J Clin Oncol. 2019;24(2):115–22.
2. Morganti S, Tarantino P, Ferraro E, D'Amico P, Duso BA, Curigliano G. Next Generation Sequencing (NGS): a revolutionary technology in pharmacogenomics and personalized medicine in cancer. Adv Exp Med Biol. 2019;1168:9–30.
3. Morganti S, Tarantino P, Ferraro E, D'Amico P, Viale G, Trapani D, et al. Complexity of genome sequencing and reporting: Next generation sequencing (NGS) technologies and implementation of precision medicine in real life. Crit Rev Oncol Hematol. 2019;133:171–82.
4. Morash M, Mitchell H, Beltran H, Elemento O, Pathak J. The role of next-generation sequencing in precision medicine: a review of outcomes in oncology. J Pers Med. 2018;8(3):30.
5. Gulilat M, Lamb T, Teft WA, Wang J, Dron JS, Robinson JF, et al. Targeted next generation sequencing as a tool for precision medicine. BMC Med Genomics. 2019;12(1):81.
6. Manzoni C, Kia DA, Vandrovcova J, Hardy J, Wood NW, Lewis PA, et al. Genome, transcriptome and proteome: the rise of omics data and their integration in biomedical sciences. Brief Bioinform. 2018;19(2):286–302.
7. Kettman JR, Frey JR, Lefkovits I. Proteome, transcriptome and genome: top down or bottom up analysis? Biomol Eng. 2001;18(5):207–12.
8. Sleator RD. The genetic code. Rewritten, revised, repurposed. Artif DNA PNA XNA. 2014;5(2): e29408.
9. Koonin EV, Novozhilov AS. Origin and evolution of the genetic code: the universal enigma. IUBMB Life. 2009;61(2):99–111.
10. Hershey JW, Sonenberg N, Mathews MB. Principles of translational control: an overview. Cold Spring Harbor PerspectBiol. 2012;4(12):a011528.

11. Pennisi E. Genomics. ENCODE project writes eulogy for junk DNA. Science. 2012;337 (6099):1159, 61.
12. Lander ES, Linton LM, Birren B, Nusbaum C, Zody MC, Baldwin J, et al. Initial sequencing and analysis of the human genome. Nature. 2001;409(6822):860–921.
13. McCombie WR, McPherson JD, Mardis ER. Next-Generation Sequencing Technologies. Cold Spring Harb Perspect Med. 2019;9(11):a036798.
14. Conesa A, Madrigal P, Tarazona S, Gomez-Cabrero D, Cervera A, McPherson A, et al. A survey of best practices for RNA-seq data analysis. Genome Biol. 2016;17:13.
15. Yan H, Tian S, Slager SL, Sun Z. ChIP-seq in studying epigenetic mechanisms of disease and promoting precision medicine: progresses and future directions. Epigenomics. 2016;8 (9):1239–58.
16. Barros-Silva D, Marques CJ, Henrique R, Jeronimo C. Profiling DNA methylation based on next-generation sequencing approaches: new insights and clinical applications. Genes (Basel). 2018;9 (9):429.
17. Lappalainen T, Scott AJ, Brandt M, Hall IM. Genomic analysis in the age of human genome sequencing. Cell. 2019;177(1):70–84.
18. Petersen BS, Fredrich B, Hoeppner MP, Ellinghaus D, Franke A. Opportunities and challenges of whole-genome and -exome sequencing. BMC Genet. 2017;18(1):14.
19. Jennings LJ, Arcila ME, Corless C, Kamel-Reid S, Lubin IM, Pfeifer J, et al. Guidelines for validation of next-generation sequencing-based oncology panels: a joint consensus recommendation of the association for molecular pathology and college of American Pathologists. J Mol Diagn. 2017;19(3):341–65.
20. Dubchak I, Poliakov A, Kislyuk A, Brudno M. Multiple whole-genome alignments without a reference organism. Genome Res. 2009;19(4):682–9.
21. de Lannoy C, de Ridder D, Risse J. The long reads ahead: de novo genome assembly using the MinION. F1000Res. 2017;6:1083.

Opportunities and Perspectives of NGS Applications in Cancer Research

2

Christian Molina-Aguilar, Martha Estefanía Vázquez-Cruz,
Rebeca Olvera-León, and Carla Daniela Robles-Espinoza

Contents

C. Molina-Aguilar · M. E. Vázquez-Cruz · R. Olvera-León
Laboratorio Internacional de Investigación sobre el Genoma Humano, Universidad Nacional
Autónoma de México, Santiago de Querétaro, México

C. D. Robles-Espinoza (✉)
Laboratorio Internacional de Investigación sobre el Genoma Humano, Universidad Nacional
Autónoma de México, Santiago de Querétaro, México

Experimental Cancer Genetics, Wellcome Sanger Institute, Cambridge, UK
e-mail: drobles@liigh.unam.mx

© Springer Nature Switzerland AG 2021   17
M. Kappelmann-Fenzl (ed.), *Next Generation Sequencing and Data Analysis*, Learning
Materials in Biosciences, https://doi.org/10.1007/978-3-030-62490-3_2

What You Will Learn in This Chapter

In this chapter, we will discuss NGS applications in cancer research, starting with a brief section on tumor driver genes and their mutational patterns, and then exploring how DNA and RNA sequencing can aid cancer diagnosis, shed light on causal agents, elucidate the biological mechanisms that participate in tumor evolution and contribute to the design of effective therapies. The technological advances that have allowed sequencing to be fast, efficient, and cost-effective have also created technical challenges, which mainly comprise the combining, categorization, comparison, and storage of large amounts of information, followed by the need for efficient analysis methodologies to extract meaningful biological information. Considering this, we will also briefly review the existing international collaborative efforts that aim to use genome and transcriptome sequencing to deepen our understanding of cancer, and will give our vision of the opportunities that this type of research offers for cancer prevention and monitoring, the challenges it still has to overcome, and perspectives for the future.

2.1   Introduction: Using Genomic Data to Understand Cancer

Cancer is a complex group of diseases that arise when mutations accumulate in cells, leading to uncontrolled cell growth, abnormal morphology, and the ability to invade surrounding tissues [1]. Therefore, it is now generally accepted that cancer is a disease of the genome. Mutations that contribute to the acquisition of these characteristics are referred to as *driver mutations*, whereas those that "hitchhike" with these are known as *passenger mutations*. These alterations can be caused by exogenous agents, such as exposure to environmental carcinogens like ultraviolet radiation, or endogenous factors, such as defects in DNA repair genes (Fig. 2.1). When mutations occur during the lifetime of a cell these are known as *somatic mutations* (as opposed to *genetic variants,* which are those present from birth).

As cancer represents the second main cause of death worldwide [2], research efforts are focusing heavily on improving early detection, elucidating the main biological mechanisms behind tumor types and identifying potential therapeutic targets. Research in all of these areas has been boosted by whole-genome and -exome sequencing of matched tumor/ normal tissue, whose analysis allows researchers to identify mutations that fuel cancer growth and that are potentially targetable. These advances have meant that clinicians and

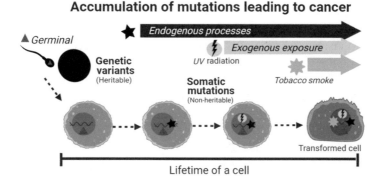

Fig. 2.1 The accumulation of mutations throughout the lifetime of a cell. These DNA lesions, caused by both exogenous and endogenous agents can lead to cancer

scientists around the world can now exploit NGS technologies and take advantage of the increasingly growing amount of sequencing information to interpret their findings. This in turn has brought us closer to precision and personalized medicine, where treatments are tailored to a particular tumor from an individual patient, potentially improving efficacy and minimizing side effects.

2.2 Driver Mutations and Their Biological Mechanisms of Action

The first genes influencing cancer development were discovered almost 40 years ago, and since then, more than 550 cancer driver genes have been described [1], most identified through modern sequencing technologies. Different types and patterns of mutations affect these genes, which are also specific to tumor stage and cancer type.

2.2.1 Oncogenes

Oncogenes are defined as those genes that have gained activity, which confers a selective advantage to the carrier cell, through the acquisition of somatic mutations. Their non-mutated counterparts are usually referred to as proto-oncogenes. Somatic gain-of-function mutations were described in the first oncogene, *RAS*, almost 40 years ago, which were shown to render the protein constitutively active and lead to continuous mitogenic signal transduction [3]. As expected, in general *gain-of-function* alterations tend to affect particular amino acids within a protein that usually lead it to be "locked in" in an active conformation by increasing its mitogenic signal transduction capabilities or by reducing its sensitivity to inhibitors.

Review Question 1

If you were studying the mutations in an oncogene identified by sequencing in a large number of tumors, what patterns of mutations would you expect to see?

To date, more than 280 genes with oncogenic activity have been described, some of which are specific to a particular cancer type. Some notable ones include *HER2* and *ESR1* in breast cancer, *BRAF* and *NRAS* in melanoma, *EGFR, ROS1, ALK*, and *RET* in lung cancer, *KRAS, BRAF, and PIK3CA* in lung and colorectal cancer, and *KIT* in acute myeloid leukemia [4]. These genes are usually affected by missense mutations, which change a single amino acid in the protein sequence, or by gene amplifications or transcriptional overexpression. Also, typically mutations in these genes are acquired somatically, as (with a few exceptions such as *RET*) germline activation of these may not be compatible with life.

2.2.2 Tumor Suppressors

Tumor suppressor genes are so named because when they become inactive through the acquisition of somatic mutations, tumor growth is accelerated. As can be expected, their functions typically involve cell cycle control, regulation of apoptosis, and DNA break repair, among others. Since inactivating mutations were found in the first discovered tumor suppressor gene, *RB1*, in 1986, more than 270 genes have been identified to contribute to cancer development when they are inactivated in different cancer types [1].

Review Question 2

If you were studying the mutations in a tumor suppressor gene identified in a large sequencing study, what patterns of mutations would you expect to see?

The most commonly mutated gene in human cancer, *TP53*, is a tumor suppressor, which regulates transcription to control cell growth arrest and apoptosis, and, therefore, the majority of cancer-derived missense mutations affect its DNA binding domain and usually are associated with advanced stages of cancer [5]. Examples of other important genes in this category are *CDKN2A* and *NF1* in melanoma, *BRCA1* and *BRCA2* in breast cancer, and *ATM* in certain leukemias [1, 4]. These genes are typically affected by stop-gain, frameshift-inducing or splice site mutations, by deletions that can span a few amino acids or the whole gene, or by gene silencing via transcriptional downregulation. Many of the genes that are cancer-predisposing in individuals are tumor suppressors, with inactivation of the remaining allele occurring later in life.

2.2.3 Gene Fusions

Gene fusions are another category of cancer-promoting genomic alterations that arise through structural rearrangements that combine two genes in a novel transcript, usually gaining a new function. One of the first cancer-promoting genetic alterations discovered was the *BCR–ABL* gene fusion, also referred to as the Philadelphia chromosome, in leukemia cells in 1959. Since then, more than 300 genes have been classified as participating in driver fusion events by either gaining oncogenic potential or by modifying their partner's function [1].

Review Question 3

When analyzing cancer RNA-Seq data, how would you identify gene fusions?

2.3   Sequencing in Cancer Diagnosis

Knowing the genes that are involved in cancer development, coupled with the ability to cost-effectively perform gene sequencing and bioinformatic analyses, is a powerful tool to screen patients at risk and aid genetic counseling.

At the moment, genetic tests can be requested by individuals from cancer-prone families to learn whether they are also at a higher risk for developing the disease. If the family carries a known mutation in a cancer gene, it can be identified through gene panels, which usually sequence a small number of known genes by hybridization followed by high-throughput sequencing (targeted sequencing) or by PCR followed by capillary sequencing. In the event that the gene panel testing returns with a negative result, the family may enter a research protocol where whole-exome or genome sequencing will be performed to attempt to identify novel cancer genes. These projects are usually research-focused (*i.e.*, no information is returned to the patient) and can increase their statistical detection power by aggregating a large number of families. However, bioinformatic analysis is key and both of these methodologies suffer from the identification of a large number of variants of uncertain significance (VUS).

VUS represent a challenge for bioinformaticians, medical professionals, and patients alike because their relationship to disease risk is unknown and therefore clinically unactionable. In order to alleviate this issue, the American College of Medical Genetics and Genomics (ACMG) published in 2000, and revised in 2007 and 2015 [6], a series of recommendations to classify variants into five categories based on population frequencies, computational predictions, functional data, and familial disease/variant co-segregation observations.

Variant classification:

- pathogenic
- likely pathogenic
- uncertain significance
- likely benign
- benign

However, as of 2019, nearly half of the half million variants in ClinVar remain in the VUS category [6]. Therefore, deep bioinformatic analyses are urgently needed to assess the increasingly large number of variants and attempt to prioritize those that may be important for disease risk and progression.

Another exciting application of high-throughput sequencing to cancer diagnosis is the analysis of liquid biopsies. Liquid biopsies refer to the non-invasive sampling of body fluids such as blood, urine, saliva, and cerebrospinal fluids, among others. The early detection of cancer in this way is theoretically possible because tumors shed cells and DNA (referred to as "circulating tumor cells," CTCs, and "circulating tumor DNA," ctDNA, respectively) into the bloodstream by apoptosis or necrosis. Several specialized techniques, such as antibody capture, depletion of white and red blood cells, and size exclusion have been applied to the detection of CTCs; cell-free DNA (cfDNA) in the blood, of which ctDNA represents between <0.1 and 10%, can be analyzed for common driver mutations [7]. This technology offers great potential, as for example, several studies have shown that *RAS-* or *TP53*-cancer associated mutations can be discovered in sputum or plasma several months before lung adenocarcinoma or bladder cancer diagnosis, respectively [5], and the size of cfDNA fragments may indicate a tumoral origin. Still, great challenges need to be overcome for this technique to be deployed for the screening of an asymptomatic population: First, the amount of ctDNA is so low that even though specialized techniques have been developed for its analysis, such as digital PCR, droplet digital PCR, and BEAM, these still suffer from low multiplexing capacity where only a few mutations can be assessed. Second, high-throughput sequencing technologies, while able to assess a large number of loci, do not have the required sensitivity to confidently detect these mutations [7]. These complications would increase the number of false-positive diagnoses, and though the technique is developing quickly, it is still at an early stage. Liquid biopsies have presently better value for prognostic assessments and disease monitoring, which we will discuss in Sect. 2.6.

2.4   Genome Sequences Can Reveal Cancer Origins

As mentioned in the introduction, mutations in the genome of a cancer cell can be caused by exogenous factors, such as exposure to carcinogenic agents like ultraviolet radiation or cigarette smoke, or endogenous processes such as defects in the DNA damage repair machinery. As these mutagenic agents have very diverse modes of action (e.g., bulky DNA adducts bind covalently to DNA bases, ionizing radiation is able to induce DNA breaks by disrupting chemical bonds, alkylating agents can add alkyl groups to guanine bases, etc.) we can expect that the patterns of mutations these leave in the genome are also quite different. For example, UV radiation preferentially causes C>T transitions at dipyrimidine sites, and exposure to benzo-[α]-pyrene results mainly in C>A mutations [8]. Therefore, the set of all mutations in a genome can be considered as an archeological record, which we can explore to learn about the mutational processes that a tumor has been exposed to (Fig. 2.2) [8].

Review Question 4

Which cancer types do you expect to have the highest and the lowest numbers of mutations, and why?

Recently, computational methodologies have been developed to extract all these mutational signatures from any given mutation catalog C, which is a matrix that has samples as columns and mutation classes as rows. The latter are the six possible mutation types (C:G > A:T, C:G > G:C, C:G > T:A, T:A > A:T, T:A > C:G, and T:A > G:C) taken into their trinucleotide contexts (*i.e.*, the base before and after each mutation), thus yielding $6 \times 4 \times 4 = 96$ different mutation classes (Fig. 2.2). These algorithms try to optimally solve $C \approx SE$, where S is the signature matrix (with mutation classes as rows and signatures as columns) and E is the exposure matrix (with samples as columns and signatures as rows) [9]. This way, we can learn which samples have which signatures with which "intensity" (the exposure).

Mutational signature analysis has been tremendously useful in recent years to elucidate the mechanism of action of several carcinogens (Fig. 2.3). For example, this type of analysis revealed in 2013 the types of mutations induced by aristolochic acid, a substance present in plants traditionally used for medicinal purposes, which are dominated by A>T transversions by formation of aristolactam-DNA adducts [10]. Another example is the discovery of the extent to which APOBEC enzymes play a role in cancer and associated cell line models by their strong activity as cytidine deaminases [11]. However, this exciting field is in constant evolution and several distinct bioinformatic methodologies have been published for the extraction of mutational signatures from cancer genomes. The original method, a *de novo* extraction algorithm based on non-negative matrix factorization (NNMF), was published by Alexandrov and collaborators and applied to 7042 distinct tumors from 30 different types of cancer, being able to identify 21 mutational signatures [8]. Since then, other methods, both *de novo* and approaches fitting C to a matrix of known signatures, have been published with varying results. Researchers doing this type of analysis can run into problems such as ambiguous signature assignment, the fact that some localized mutational processes may not be taken into account, and the algorithms' assumption that all samples being analyzed have a similar mutational profile [9]. A recent study has suggested that a combination of *de novo* and fitting approaches may reduce false positives while still allowing the discovery of novel signatures [9]. Generally, as is the case with any bioinformatic tool, researchers must be cautious about their results and should perform a manual curation where possible, making use of prior biological knowledge and making sure results make sense.

Mutational signature analysis has recently been expanded to include multiple-base mutations and small indels. To date, nearly 24,000 cancer genomes and exomes have been analyzed, with 49 single-base substitution (SBS), 11 doublet-base substitutions, four clustered base substitutions, and 17 indel mutational signatures discovered [8].

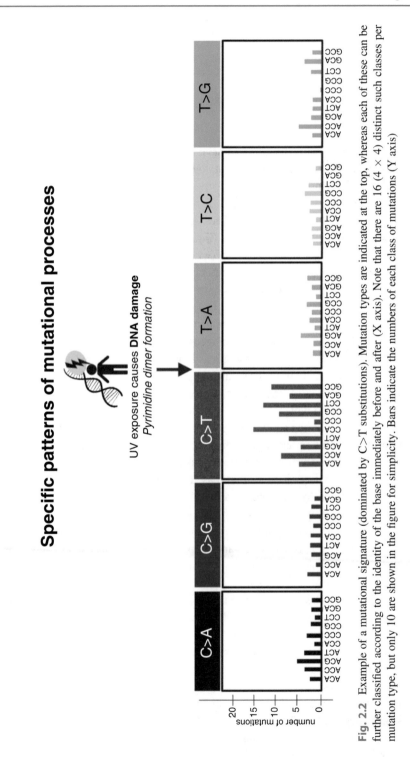

Fig. 2.2 Example of a mutational signature (dominated by C>T substitutions). Mutation types are indicated at the top, whereas each of these can be further classified according to the identity of the base immediately before and after (X axis). Note that there are 16 (4 × 4) distinct such classes per mutation type, but only 10 are shown in the figure for simplicity. Bars indicate the numbers of each class of mutations (Y axis)

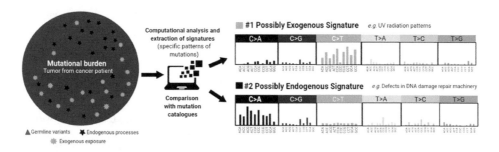

Fig. 2.3 Computational analysis of tumor mutational burden can elucidate the contribution of mutational processes. The genome sequence of tumors can be analyzed by sophisticated algorithms to extract mutational signatures (in this case, the two shown at the right), which then can be compared with known catalogs to assign a potential etiological process (in this case, UV radiation may be causing the first signature, whereas the second one may be caused by damage to the DNA repair machinery). These processes then have been operational during the lifetime of the tumor and have contributed to its growth

2.5 Genome and Transcriptome Sequences Are Useful for Elucidating Cancer Biology

Genome and transcriptome sequencing have allowed researchers to gain unprecedented resolution into the cellular processes that lead to cell transformation and the consequences of carrying particular mutations. Traditionally, scientists have relied on "bulk" sequencing, which means that they sequence the genomes or transcriptomes of a bulk of cells from the tumor and therefore the experiment readout can be interpreted as an average of the tumor mutations or gene expression patterns. One of the earliest large cancer sequencing projects, The Cancer Genome Atlas (TCGA, discussed below in Sect. 2.7), bulk-sequenced exomes and transcriptomes from large numbers and types of cancer and allowed the identification of further genomic drivers, the classification of several types of cancer in genomic subtypes, and the discovery of dysregulated processes leading to cell growth and metastasis [12]. However, many fundamental discoveries in cancer genomics have come from the study of in vitro and in vivo models of the disease.

2.5.1 Sequencing of In Vitro and In Vivo Tumor Models Identifies Fundamental Biological Properties

The first available models for studying cancer cell biology were cell lines, an *in vitro* strategy fueled in 1951 with the isolation and culture of biopsy cells taken from cervical cancer patient Henrietta Lacks (who did not consent to the experiment) [13]. After these cells (termed HeLa, after the patient's name) proved invaluable for, *e.g.*, developing the polio vaccine and defining the effects of X-ray radiation on human cells [14], researchers realized their value and set out to derive cell lines from lung, breast, ovarian, colon, prostate, skin, renal, CNS, and other tissues [15]. There are now more than 1,300 cancer cell lines [15], which are still routinely used by scientists around the world. Sequencing of these has revealed the extent to which cell lines resemble the human tumor, aiding model selection, and has described the mutational signatures present in human tumors as well as their dynamics [11]. However, recent criticism has focused on the fact that the majority of these cell lines are derived from European-descent populations, which may mean that disease dynamics in other genetic ancestries may not be well-modeled by these cells [15].

Another *in vitro* model recently developed are 3D organoids [16]. Organoids are "artificially grown masses of cells or tissue that resemble an organ" [17], and have been found to faithfully reproduce the original tumors. Particularly, in an early effort to generate a "living biobank" of organoids from colorectal cancer (CRC) patients, exome sequencing found that organoids have a similar mutation bias (CpG > T transitions) to that of earlier, large-scale CRC sequencing efforts, as well as having a similar frequency of hypermutated tumors and maintaining common driver mutations [18]. RNA analysis of the same organoids identified differential expression of cancer-associated genes such as *PROX1* and *PTCH1*, as well as having similar expression profiles to other CRC tumors. Follow-up studies on lung, esophagus, pancreas, and other tissues have shown similar results [16], which demonstrates the power of this system to model cancer evolution. These have been used successfully to investigate the link between infectious agents and cancer [16], and to describe the mutational signatures in healthy stem cells that lead to malignant transformation [19]. Nevertheless, these are still not fully reproducible, showing a great deal of variation, limiting their current applicability [16].

In vivo models have also been used to explore cancer biology in the context of a full host organism. Particularly, patient-derived xenografts (PDXs) have emerged as a faithful preclinical model to recapitulate tumor histology, genome, transcriptome, and heterogeneity, as well as drug response. These are models in which fresh tumor tissue is directly transplanted into immunocompromised rats or mice either subcutaneously or orthotopically [20]. Whole-exome sequencing of PDX models has been used to identify targetable genomic alterations, and their transcriptomic characterization at the single-cell level has identified subpopulations of cells that provide drug resistance in melanoma [20]. Large repositories, such as the one in The Jackson Laboratory, which comprises 455 PDX models from more than 30 primary sites, are being genomically and transcriptomically characterized to maximize their utility for translational studies [21]. Even though their utility in preclinical research has been extensively recognized, PDX model generation requires high technical skills and can have low success rates, and can take several months to establish [20], which represent important limitations. Additionally, as the host animals are immunodeficient, a different model would need to be chosen for the study of tumor–immune cell interactions, a very important contributor to tumor dynamics.

2.5.2 Single-Cell DNA and RNA Sequencing

More recently, advances in fields such as microfluidics and nanotechnology have resulted in a number of methodologies for assessing genome sequence, messenger RNA levels, protein abundance, and chromatin accessibility at the single-cell level [22]. For the cancer field, these exciting developments have meant that the cell subpopulations conforming tumors have become apparent, analyses have revealed rare cell subtypes, and they have also helped pinpoint which ones of these are able to re-establish growth after treatment [22].

One of the most exciting observations stemming from these types of analyses is the discovery that many of the mutations in a tumor are subclonal, therefore, meaning that different sections within the same tumor may have different genomic alterations [22]—therefore adding to the notion that a single biopsy may not be enough to identify all tumor drivers. As another example of the power of these studies, researchers based at the Broad Institute of MIT and Harvard identified, in melanoma tumors, a cellular program associated with immune evasion and T-cell exclusion that is present from before therapy, and that is able to predict responses to this type of treatment in an independent cohort of patients [23]. But perhaps the most striking discovery has been the identification of non-genetic mechanisms underlying the emergence of drug treatment resistance, which we will discuss in detail in Sect. 2.6.

Although the number of methodologies published for single-cell DNA and RNA analysis has been growing steadily over the last few years, the field still has numerous challenges to overcome: The development of methods to efficiently and non-disruptively isolate single cells from the tissue of origin, the amplification of that individual cell's biological material for downstream processing, and the subsequent analysis of that material to identify the variation of interest as well as taking into account the potential errors and biases [24]. However, these methodologies are set to become standard in the near future, as large consortia such as The Human Cell Atlas are already undertaking a number of studies to fulfill their aim of defining all human cell types at the molecular and morphological levels by sequencing at least 10 billion single cells from healthy tissues [25]. It is not difficult to envision, thus, that the technological advances stemming from this and related projects may be exploited in a large follow-up project to TCGA where the focus will be characterizing individual cells, the abundances of cell states in distinct tumors, and the identification of therapy-resistant subclones.

2.5.3 Exploring Intra-Tumor Heterogeneity Through Sequencing

It has also been recognized that tumors are highly heterogeneous masses of distinct cell types, clonal and subclonal genomic mutations as well as a mixture of distinct transcriptional cell states [26]. Bulk DNA and RNA sequencing can be exploited to investigate this heterogeneity by specialized bioinformatics analyses. One early study using bulk genome sequencing analyzed 21 breast tumors and concluded, via the study of mutations that are present in only a small fraction of the sequencing reads, that all samples had more subclonal than clonal mutations and authors were able to reconstruct phylogenetic trees for some of the tumors [27]. Since then, many other studies have been published where similar analyses have informed our understanding of tumor evolution, selection dynamics, and mutation cooperation [26]. This approach demonstrates the power of careful analysis of sequencing reads and reveals the amount of information that can be inferred from these experiments.

Bulk transcriptome sequencing can also be examined beyond the differential expression analyses that are typically done with such data. A number of different methodologies have been developed to perform "cell type deconvolution," a methodology used to infer the proportions of different cell types in a tissue sample using computational approaches based on specific marker genes or expression signatures. Deconvolution methods quantitatively estimate the fractions of individual cell types in a heterocellular tissue (such as the tumor microenvironment) by considering the bulk transcriptome as the "convolution" of cell-specific signatures [28]. One of the most popular method is CIBERSORT [28], which is able to estimate the immune cell fraction component of a tumor biopsy. This algorithm has been used to identify immune infiltration signatures in distinct cancer types and their relationship to survival patterns and other clinical characteristics [29]. Other similar methods are MuSiC, deconvSeq, and SCDC [30]. The choice of method would depend on the type of tissue being studied and the biological question being addressed.

Of course, the exploration of intra-tumor heterogeneity has been spectacularly boosted by the development of single-cell sequencing technologies. Recent studies exploiting this technology have been able to show that inter-tumor heterogeneity in cancer cells is much larger than intra-tumor heterogeneity, whereas this is not the case for non-malignant cells, and to dissect patterns of heterogeneity (such as cell cycle stage and hypoxia response) from context-specific programs that determine tumor progression and drug response [31]. These advances would have been impossible without the ability to discern cell types and cell states within a tumor and have greatly informed our understanding of tumor dynamics.

2.6   Sequencing in Cancer Treatment

Perhaps the area in which tumor sequencing has had a more tangible impact has been in precision and personalized treatment design. Whole-genome and -exome sequencing of large groups of tumors has made it possible to identify genomic and transcriptomic subtypes in neoplasia from tissues such as breast, skin, and bladder, among others [32]. These genomic subtypes are associated with different clinical presentations and molecular characteristics, and can be targeted with different treatments. For example, in melanoma, four genomic subtypes have been identified: Tumors that have *BRAF* mutated, almost always at the V600 residue, which present at a younger age and can be targeted with BRAF inhibitors such as vemurafenib, those that have a *RAS* gene mutated, characterized by hyperactivation of the MAPK pathway and can perhaps be targeted with *BET* and *MEK* inhibitors, those with loss of the tumor suppressor *NF1*, which have a higher mutational burden and present at an older age, and those without mutations in any of these driver genes [33]. In colon cancer, different genomic subtypes can be recognized with chromosomal unstable tumors having defects in chromosome segregation, telomere stability, and DNA damage response and hypermutated tumors usually having a defective DNA mismatch repair system [34]. The latter respond well to immune checkpoint inhibitors, especially if they have a high mutational burden, and, therefore, having a hypermutator phenotype is also a biomarker that can help stratify patients for treatment choice.

Transcriptomic characterization of tumors has also been of great use to identify subtypes within a cancer type that can be treated specifically to boost therapy efficacy. For example, breast cancer has been classified into subtypes according to gene expression: Luminal A, luminal B, *ERBB2*-overexpressing, normal-like, and basal-like [35]. This classification takes into account expression of the estrogen receptor (ER+), the progesterone receptor (PR +), and *ERBB2* amplification [35]. Luminal tumors are likely to respond to endocrine therapy, while those with *ERBB2* amplification can be targeted with trastuzumab and chemotherapy. Treatment of triple-negative breast cancers is more challenging, but PARP inhibitors and immunotherapy are beginning to be tested in the clinic [36]. Melanoma tumors have also been classified into expression subtypes with potential therapeutic implications [37]. Therefore, the sequencing of both tumor DNA and RNA can inform about the biology of cancer and help identify potential therapeutic targets.

As briefly mentioned above, the most useful aspects of tumor sequencing are the identification of therapeutical targets and the discovery of tumor biomarkers. Apart from classical such biomarkers like the presence of established driver mutations, another one that has emerged as an important predictor of response to immunotherapy is tumor mutational burden (TMB), as some authors suggest that higher TMB predicts favorable outcome to PD-1/PD-L1 blockade across diverse tumors [38]. Tumors with high TMB are more likely to respond to this type of treatment because the number of "neoepitopes" is higher, this is, the amount of novel peptides that arise from tumor-specific mutations and can be recognized by the immune system [36]. Chromosomal instability, which can also be detected by sequencing, may also impact therapeutic response.

An exciting development fueled by the ability to identify neoepitopes by next-generation sequencing is adoptive T-cell therapy. In this form of treatment, tumor/normal sample pairs are sequenced in order to identify novel mutations that may be targetable by the immune system [39]. After these are validated by RNA expression analysis and mass spectrometry, and are deemed good ligands for HLA molecules by bioinformatic analyses, then they are co-cultured with tumor-infiltrating lymphocytes (TILs) resected from a tumor biopsy from the same patient [40]. This methodology then allows the selective expansion of TILs with a specific reactivity to the target tumor, and can then be introduced back into the patient.

Another topic that has gained traction in recent years due to the advent of single-cell transcriptome sequencing is the realization that drug resistance in cancer can be generated by both genetic and non-genetic mechanisms. Genetic resistance appears when mutations emerge that allow the cell to grow even in the presence of drug. Examples of this are resistance to *BRAF* inhibitors by amplification of *BRAF* or mutations of *NRAS* in melanoma [41], and hormone therapy resistance in breast cancer by mutation of the ER [42]. However, non-genetic mechanisms can also mediate resistance to targeted therapy. In a recent study, Rambow and colleagues elegantly demonstrated that *BRAF* and *MEK* inhibition in a PDX model of melanoma resulted in the establishment of minimal residual disease (MRD), whose mutational profile was not significantly different to that of the tumor before treatment [22]. Furthermore, cells in MRD could be classified in four transcriptional states, one of which, characterized by having a neural crest stem cell transcriptional program, went on to establish tumor growth despite continuous treatment. This subpopulation could be further targeted by a retinoid X receptor inhibitor, delaying the onset of resistance, and demonstrating the power of these kinds of analyses to identify potential targets beyond the findings from bulk genome and transcriptome sequencing.

Liquid biopsies are also useful for the monitoring of cancer evolution and progression, as they can inform in real time how the tumor is responding to therapy and whether novel mutations have been acquired that may provide resistance. The tumor mutational landscape changes over time due to evolutionary and therapeutic selective pressure [26], thus, almost all tumors acquire resistance to systemic treatment as a result of tumor heterogeneity, clonal evolution, and selection. For example, in a study of 640 patients, Bettegowda and colleagues found ctDNA fragments at relatively high concentrations in the circulation of most patients with metastatic cancer and at lower fraction of patients with localized cancers, as well as identifying mutations in ctDNA that conferred resistance to EGFR blockade in colorectal cancer patients [43]. These and other similar results illustrate the potential of this technology to aid in monitoring tumor evolution and therapy treatment.

2.7 International Collaborative Efforts in Cancer Sequencing and Mutation Classification

In order to exploit the power of next-generation sequencing technologies in cancer diagnosis, monitoring, and treatment, international collaborative consortia have been formed to collect and sequence DNA and RNA of thousands of cancer tissues from different countries and research institutions around the world.

2.7.1 The Cancer Genome Atlas (TCGA)

TCGA was launched in 2005 as an effort to generate sequencing data from a large collection of tumors, as well as to analyze and interpret their molecular profiles to provide a comprehensive overview of the underlying biology and potential therapeutic targets [12]. Other secondary aims of this project are to release data freely to the scientific community, to train expert individuals, and to develop specialized computational tools and infrastructure that may be useful to other large-scale projects. As such, many secondary analyses have allowed deep exploration of these data beyond the primary reports released by TCGA Network. In 2018, TCGA published the Pan-Cancer Atlas, a number of coordinated papers reporting on the analysis of over 11,000 tumors from 33 cancer type [44]. Data access website: https://portal.gdc.cancer.gov/.

2.7.2 International Cancer Genome Consortium (ICGC)

ICGC is a voluntary, international initiative launched in 2007 to coordinate the analysis of tumor genomes from 25 primary untreated tumors from 50 cancer types from around the world [45]. In 2019, its portal contained data aggregated from more than 20,000 contributors, including TCGA, and had information on about 77 million somatic mutations [46]. ICGC has numerous initiatives, including PCAWG (reviewed below), The ICGC for Medicine Initiative (ICGCmed), and ICGC for Accelerating Research in Genomic Oncology (ICGC-ARGO). Website: https://icgc.org/.

2.7.3 Pan-Cancer Analysis of Whole Genomes (PCAWG)

PCAWG is an initiative from ICGC that aims to study more than 2,600 cancer whole genomes from 39 distinct tumor types, aiming to be a follow-up analysis to those performed on coding sequences [47]. The PCAWG Network's first publications, published in 2020, focus on cataloging non-coding driver mutations, tumor evolutionary history, identifying structural variation, mutational signatures analysis and inferring interactions between somatic and germline variation [47]. ICGC has made data on somatic calls, including SNPs, indels, structural and copy number variants, available for use by any researcher. Website: https://dcc.icgc.org/pcawg.

2.7.4 Catalog of Somatic Mutations in Cancer (COSMIC)

COSMIC was launched in 2004 as a database of information on somatic mutations in human cancer [48]. It has steadily grown over the years and in its latest release to this date (in September 2019) it had information on every human gene, and on nearly nine million coding mutations from more than 1.4 million samples [48]. Data in COSMIC is manually curated by experts constantly reviewing the literature, as well as from systematic screens released with these publications. It is widely used for exploring available information on frequency and potential pathogenic consequences of somatic mutations. Website: https://cancer.sanger.ac.uk/cosmic.

2.7.5   ClinVar

ClinVar was released in 2013 as a freely available resource that catalogs genome variation of clinical importance, incorporating information on the genomic variants, the submitter, the associated phenotype, the clinical interpretation, and the supporting evidence [49]. Terminology for variant interpretation follows the recommendations by the ACMG (reviewed above). All data have been made available for use by researchers in multiple formats, and, therefore, has become a valuable database for aggregating and consulting medically important genome variation. Website: http://www.ncbi.nlm.nih.gov/clinvar/.

2.8   Opportunities, Challenges, and Perspectives

There is no doubt that genomics is already playing a large role in cancer diagnosis and treatment, but it may become even more important in the near future. An ideal scenario to treat a patient with cancer would be to have all possible information at hand before treatment choice, which includes whole genome sequencing. In fact, the United Kingdom through its National Health System is already setting up plans to whole-genome sequence every child with cancer as well as sequencing a large part of their patient and healthy population through the 100,000 Genomes Project [50] and the UK Biobank [51]. Similarly, the United States under the Obama administration announced the Precision Medicine Initiative in 2015, funding the National Health Institutes to form a cohort of a million volunteers to provide genomic data and medical records, among others [52]. These programs illustrate that policy-makers recognize the power that this technology can bring to the clinic and are working already to make it a reality. People also recognize the benefits that knowing their genome sequence can bring them, evidenced by the fact that the number of humans around the world estimated to have been sequenced has dramatically increased from one in 2003 to over 1.5 million in 2018 [53].

The promise of precision and personalized genomic medicine is exciting and potentially life-changing, and it has already revolutionized the fields of rare disease diagnosis by identifying causal mutations in a quarter of patients with a potential genetic condition [53] and non-invasive prenatal testing by allowing rapid assessment of fetal chromosomal aneuploidies [54]. The cancer field is no exception. As we have discussed throughout this Chapter, genomic approaches have greatly advanced diagnosis via genetic testing for at-risk families, disease monitoring via solid and liquid biopsy analysis and treatment through the identification of therapeutic targets, and the development of immune therapies. However, there are many challenges still to overcome in this field: To begin with, we only know the consequences of a very small number of genetic alterations, out of all possible aberrations, that may occur in a tumor. This means that even if we develop the technology to detect them with high accuracy, we still may not know whether they play an important role in disease development or are just passengers. To alleviate this, scientists are continuously working on identifying the characteristics of cancer drivers and testing potential therapeutics. Then, we also have the issue of balancing the sensitivity and the specificity of these technologies if we plan to routinely use them to screen an asymptomatic population for early signs of cancer and perform tumor monitoring, which are the goals of the liquid biopsy revolution. Continuous efforts in protocol improvement and algorithm development should bring us closer to making this a reality.

Of course, the implementation of a "genomic medicine for all" approach also carries ethical issues that need to be carefully considered. For example, every whole-genome sequencing of a cancer patient may also identify secondary findings unrelated to the disease of interest, for example, a genetic variant that increases risk to Alzheimer's disease in a breast cancer patient. Discussions are still on-going as to whether to report these findings to the patients. Another potential area of consideration is privacy and security of the patient's genetic information and its implications to the patient's relatives, as well as the definition of "informed consent" [55].

Finally, we must not forget that even though genomics is set to revolutionize our knowledge of cancer biology and the clinical care of cancer patients, it can still tell us only part of the story. Other fields, such as immunohistochemistry, histopathology, radiation oncology, imaging procedures in radiology, and molecular biology are equally necessary if we want to have a holistic view of the chain of events necessary for cancer development as well as to provide the best care to patients. There is no doubt that an interdisciplinary approach will help researchers and clinicians deliver on the promise of precision medicine.

Take Home Message
- Genes are drivers if their somatic mutation aids cancer growth. These can be classified into oncogenes (genes whose activation leads to tumor development), tumor suppressors (genes whose inactivation supports neoplastic transformation), and gene fusions (a product of two genes that gains a novel ability). These genes carry different mutation patterns.
- Sequencing has aided cancer diagnosis by facilitating gene panel testing and whole-exome/genome sequencing.
- Mutational signatures are patterns of mutations that can be extracted via computational analysis of large cohorts of tumors, and which can be informative about the processes that gave rise to a tumor.
- Bulk genome sequencing of large numbers of tumors has allowed the identification of mutational drivers, the classification of tumors in genomic subtypes, and revealed dysregulated processes critical for tumor growth.
- Single-cell DNA and RNA sequencing of cancers can reveal their biological complexity at an unprecedented level, examples include the amount of intratumor heterogeneity and drug resistance mechanisms.

(continued)

- Genome sequencing is becoming paramount in cancer treatment, allowing the identification of druggable targets and biomarkers for immunotherapy response and aiding the development of novel treatment strategies.

Answer to Question 1

Oncogenes are characterized by a hotspot pattern of mutations, in which alterations tend to cluster in a few amino acids that render the protein constitutively active. Therefore, the pattern would resemble skyscrapers. For example, the *RAS* oncogene typically harbors mutations in amino acids 12, 13, and 61 (Fig. 2.4) in human cancers, whereas *BRAF* is recurrently mutated at amino acid 600.

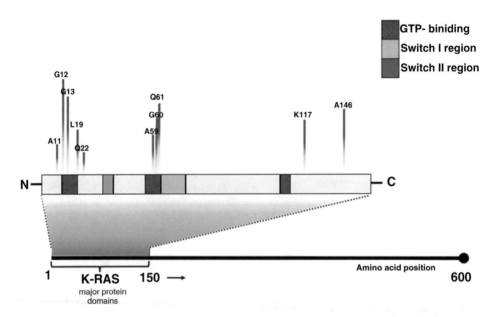

Fig. 2.4 Hotspots in the major protein domains of KRAS protein structure. Although the protein KRAS is made of 600 amino acids, the hotspots reported to date concentrate in the 150 immediately adjacent to the N-terminal end. In red, GTP-binding domains, in green, the switch I region, and in blue the switch II region. Shown are the 10 most commonly mutated amino acids in the protein, which are usually found closer to the N-terminal region. Data comprise a set of 24,592 tumor samples (from more than 30 different tumor types). G12 was mutated in 2175 samples, G13 in 264 samples, and Q61 in 190 samples. Data obtained in December 2019 from the Cancer Hotspot resource (Memorial Sloan Kettering Cancer Center, http://www.cancerhotspots.org/#/home)

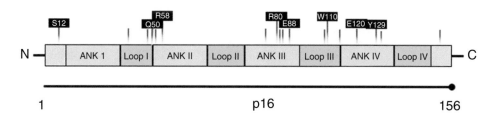

Fig. 2.5 Mutated residues in cancer samples in the p16 protein structure. Blue and yellow boxes indicate distinct domains. Red lines indicate changes of a single residue. Those that lead to a stop-gain are indicated in a black box. Data obtained in December 2019 from the Cancer Hotspot resource. Ankyrin-binding domains (ANK) (Memorial Sloan Kettering Cancer Center, http://www.cancerhotspots.org/#/home)

Answer to Question 2

Tumor suppressors are characterized by an evenly distributed pattern of inactivating mutations, so their mutational pattern would more or less look flat. However, the key here is that a large fraction of mutations tending to be stop-gain, frameshift-inducing, or to affect splice sites. For example, *p16* (one of the proteins encoded by the *CDKN2A* gene) consists of 156 amino acids, of which 19 have been found to be affected by somatic mutations scattered throughout the protein (Fig. 2.5).

It is important to highlight that eight of the mutations (black boxes in Fig. 2.5) induce a change of single residue causing the formation of non-functional truncated proteins.

Answer to Question 3

Fusion events can be discovered through the analysis of "split reads," which are those that map to two different genes in the human reference genome. If a fusion event is consistently observed in different samples, this becomes a candidate to be a cancer driver.

Answer to Question 4

Typically, the types of cancer with the highest numbers of mutations are those with a strong environmental component, such as melanoma (UV light exposure) or lung cancer (cigarette smoke) because individuals can be highly exposed to these mutagens for long periods of time. The types of cancer with the lowest numbers of mutations are those that appear in childhood, such as hepatoblastoma and pilocytic astrocytoma, due to a lower exposure to mutagens and/or endogenous mutations.

Acknowledgements We thank Dr. Stefan Fischer (Biochemist at the Faculty of Applied Informatics, Deggendorf Institute of Technology, Germany) for reviewing this chapter and suggesting extremely relevant enhancements to the original manuscript. The authors wish to thank Jair S. García-Sotelo, Alejandro de León, Carlos S. Flores, and Luis A. Aguilar of the Laboratorio

Nacional de Visualización Científica Avanzada from the National Autonomous University of Mexico, and Alejandra Castillo, Carina Díaz, Abigayl Hernández and Eglee Lomelin of the International Laboratory for Human Genome Research, UNAM.

References

1. Sondka Z, Bamford S, Cole CG, Ward SA, Dunham I, Forbes SA. The COSMIC cancer gene census: describing genetic dysfunction across all human cancers. Nat Rev Cancer. 2018;18 (11):696–705.
2. The International Agency for Research on Cancer (IARC). Latest global cancer data: Cancer burden rises to 18.1 million new cases and 9.6 million cancer deaths in 2018. p. https://www.who.int/cancer/PRGlobocanFinal.pdf.
3. Campbell PM, Der CJ. Oncogenic Ras and its role in tumor cell invasion and metastasis. Semin Cancer Biol. 2004;14(2):105–14.
4. Liu Y, Sun J, Zhao M. ONGene: a literature-based database for human oncogenes. J Genet Genomics. 2017;44(2):119–21.
5. Gormally E, Vineis P, Matullo G, Veglia F, Caboux E, Le Roux E, et al. TP53 and KRAS2 mutations in plasma DNA of healthy subjects and subsequent cancer occurrence: a prospective study. Cancer Res. 2006;66(13):6871–6.
6. Pérez-Palma E, Gramm M, Nürnberg P, May P, Lal D. Simple ClinVar: an interactive web server to explore and retrieve gene and disease variants aggregated in ClinVar database. Nucleic Acids Res. 2019;47(W1):W99–105.
7. Chen M, Zhao H. Next-generation sequencing in liquid biopsy: cancer screening and early detection. Hum Genomics. 2019;13(1):34.
8. Alexandrov LB, Kim J, Haradhvala NJ, Huang MN, Ng AW, Wu Y, et al. The repertoire of mutational signatures in human cancer. bioRxiv. 2019:322859.
9. Maura F, Degasperi A, Nadeu F, Leongamornlert D, Davies H, Moore L, et al. A practical guide for mutational signature analysis in hematological malignancies. Nat Commun. 2019;10(1):2969.
10. Hoang ML, Chen C-H, Sidorenko VS, He J, Dickman KG, Yun BH, et al. Mutational signature of aristolochic acid exposure as revealed by whole-exome sequencing. Sci Transl Med. 2013;5 (197):197ra102.
11. Petljak M, Alexandrov LB, Brammeld JS, Price S, Wedge DC, Grossmann S, et al. Characterizing mutational signatures in human cancer cell lines reveals episodic APOBEC mutagenesis. Cell. 2019;176(6):1282–1294.e20.
12. Cancer Genome Atlas Research Network, Weinstein JN, Collisson EA, Mills GB, KRM S, Ozenberger BA, et al. The Cancer Genome Atlas Pan-Cancer analysis project. Nat Genet. 2013;45(10):1113–20.
13. Lucey BP, Nelson-Rees WA, Hutchins GM. Henrietta Lacks, HeLa cells, and cell culture contamination. Arch Pathol Lab Med. 2009;133(9):1463–7.
14. Puck TT, Marcus PI. Action of x-rays on mammalian cells. J Exp Med. 1956;103(5):653–66.
15. Dutil J, Chen Z, Monteiro AN, Teer JK, Eschrich SA. An interactive resource to probe genetic diversity and estimated ancestry in cancer cell lines. Cancer Res. 2019;79(7):1263–73.
16. Drost J, Clevers H. Organoids in cancer research. Nat Rev Cancer. 2018;18(7):407–18.
17. Organoid | Definition of Organoid by Lexico [Internet]. [cited 2019 Dec 19]. Retrieved from: https://www.lexico.com/en/definition/organoid.
18. van de Wetering M, Francies HE, Francis JM, Bounova G, Iorio F, Pronk A, et al. Prospective derivation of a living organoid biobank of colorectal cancer patients. Cell. 2015;161(4):933–45.

19. Blokzijl F, de Ligt J, Jager M, Sasselli V, Roerink S, Sasaki N, et al. Tissue-specific mutation accumulation in human adult stem cells during life. Nature. 2016;538(7624):260–4.
20. Woo XY, Srivastava A, Graber JH, Yadav V, Sarsani VK, Simons A, et al. Genomic data analysis workflows for tumors from patient-derived xenografts (PDXs): challenges and guidelines. BMC Med Genomics. 2019;12(1):92.
21. Stuart T, Satija R. Integrative single-cell analysis. Nat Rev Genet. 2019;20(5):257–72.
22. Rambow F, Rogiers A, Marin-Bejar O, Aibar S, Femel J, Dewaele M, et al. Toward minimal residual disease-directed therapy in melanoma. Cell. 2018;174(4):843–855.e19.
23. Jerby-Arnon L, Shah P, Cuoco MS, Rodman C, Su M-J, Melms JC, et al. a cancer cell program promotes T cell exclusion and resistance to checkpoint blockade. Cell. 2018;175(4):984–997.e24.
24. Gawad C, Koh W, Quake SR. Single-cell genome sequencing: current state of the science. Nat Rev Genet. 2016;17(3):175–88.
25. Regev A, Teichmann SA, Lander ES, Amit I, Benoist C, Birney E, et al. The human cell atlas. eLife. 2017;05(6)
26. Ben-David U, Beroukhim R, Golub TR. Genomic evolution of cancer models: perils and opportunities. Nat Rev Cancer. 2019;19(2):97–109.
27. Nik-Zainal S, Van Loo P, Wedge DC, Alexandrov LB, Greenman CD, Lau KW, et al. The life history of 21 breast cancers. Cell. 2012;149(5):994–1007.
28. Chen B, Khodadoust MS, Liu CL, Newman AM, Alizadeh AA. Profiling tumor infiltrating immune cells with CIBERSORT. Methods Mol Biol Clifton NJ. 2018;1711:243–59.
29. Thorsson V, Gibbs DL, Brown SD, Wolf D, Bortone DS, Ou Yang T-H, et al. The immune landscape of cancer. Immunity. 2018;48(4):812–830.e14.
30. Cobos FA, Alquicira-Hernandez J, Powell J, Mestdagh P, Preter KD. Comprehensive benchmarking of computational deconvolution of transcriptomics data. bioRxiv. 2020; 2020.01.10.897116.
31. Suvà ML, Tirosh I. Single-cell RNA sequencing in cancer: lessons learned and emerging challenges. Mol Cell. 2019;75(1):7–12.
32. Hurst CD, Alder O, Platt FM, Droop A, Stead LF, Burns JE, et al. Genomic subtypes of non-invasive bladder cancer with distinct metabolic profile, clinical outcome and female gender bias in KDM6A mutation frequency. Cancer Cell. 2017;32(5):701–715.e7.
33. Echevarría-Vargas IM, Reyes-Uribe PI, Guterres AN, Yin X, Kossenkov AV, Liu Q, et al. Co-targeting BET and MEK as salvage therapy for MAPK and checkpoint inhibitor-resistant melanoma. EMBO Mol Med. 2018;10:5.
34. Dienstmann R, Vermeulen L, Guinney J, Kopetz S, Tejpar S, Tabernero J. Consensus molecular subtypes and the evolution of precision medicine in colorectal cancer. Nat Rev Cancer. 2017;17 (2):79–92.
35. Fragomeni SM, Sciallis A, Jeruss JS. Molecular subtypes and local-regional control of breast cancer. Surg Oncol Clin N Am. 2018;27(1):95–120.
36. Jiang Y-Z, Ma D, Suo C, Shi J, Xue M, Hu X, et al. Genomic and transcriptomic landscape of triple-negative breast cancers: subtypes and treatment strategies. Cancer Cell. 2019;35 (3):428–440.e5.
37. Jönsson G, Busch C, Knappskog S, Geisler J, Miletic H, Ringnér M, et al. Gene expression profiling-based identification of molecular subtypes in stage IV melanomas with different clinical outcome. Clin Cancer Res Off J Am Assoc Cancer Res. 2010;16(13):3356–67.
38. Goodman AM, Kato S, Bazhenova L, Patel SP, Frampton GM, Miller V, et al. Tumor mutational burden as an independent predictor of response to immunotherapy in diverse cancers. Mol Cancer Ther. 2017;16(11):2598–608.
39. Ott PA, Dotti G, Yee C, Goff SL. An update on adoptive T-cell therapy and neoantigen vaccines. Am Soc Clin Oncol Educ Book Am Soc Clin Oncol Annu Meet. 2019;(39):e70–8.

40. Liu XS, Mardis ER. Applications of Immunogenomics to Cancer. Cell. 2017 09;168(4):600–12.

41. Manzano JL, Layos L, Bugés C, de los Llanos Gil M, Vila L, Martínez-Balibrea E, et al. Resistant mechanisms to BRAF inhibitors in melanoma. Ann Transl Med. 2016;4(12). Available from: https://www.ncbi.nlm.nih.gov/pmc/articles/PMC4930524/

42. AlFakeeh A, Brezden-Masley C. Overcoming endocrine resistance in hormone receptor-positive breast cancer. Curr Oncol Tor Ont. 2018;25(Suppl 1):S18–27.

43. Bettegowda C, Sausen M, Leary RJ, Kinde I, Wang Y, Agrawal N, et al. Detection of circulating tumor DNA in early- and late-stage human malignancies. Sci Transl Med. 2014;6(224):224ra24.

44. Ding L, Bailey MH, Porta-Pardo E, Thorsson V, Colaprico A, Bertrand D, et al. Perspective on oncogenic processes at the end of the beginning of cancer genomics. Cell. 2018;173(2):305–320. e10.

45. International Cancer Genome Consortium, Hudson TJ, Anderson W, Artez A, Barker AD, Bell C, et al. International network of cancer genome projects. Nature. 2010;464(7291):993–8.

46. Zhang J, Bajari R, Andric D, Gerthoffert F, Lepsa A, Nahal-Bose H, et al. The International Cancer Genome Consortium Data Portal. Nat Biotechnol. 2019;37(4):367–9.

47. Campbell PJ, Getz G, Stuart JM, Korbel JO, Stein LD. Pan-cancer analysis of whole genomes. bioRxiv. 2017:162784.

48. Tate JG, Bamford S, Jubb HC, Sondka Z, Beare DM, Bindal N, et al. COSMIC: the Catalogue Of Somatic Mutations In Cancer. Nucleic Acids Res. 2019;47(D1):D941–7.

49. Landrum MJ, Lee JM, Riley GR, Jang W, Rubinstein WS, Church DM, et al. ClinVar: public archive of relationships among sequence variation and human phenotype. Nucleic Acids Res. 2014;42(Database issue):D980–5.

50. Turnbull C, Scott RH, Thomas E, Jones L, Murugaesu N, Pretty FB, et al. The 100 000 genomes project: bringing whole genome sequencing to the NHS. BMJ. 2018;24(361):k1687.

51. biobank [Internet]. UK Biobank leads the way in genetics research to tackle chronic diseases. 2019. Available from: https://www.ukbiobank.ac.uk/2019/09/uk-biobank-leads-the-way-in-genetics-research-to-tackle-chronic-diseases/.

52. Office of the Press Secretary. The White House [Internet]. FACT SHEET: President Obama's Precision Medicine Initiative. 2015. Available from: https://obamawhitehouse.archives.gov/the-press-office/2015/01/30/fact-sheet-president-obama-s-precision-medicine-initiative

53. Shendure J, Findlay GM, Snyder MW. Genomic medicine-progress, pitfalls, and promise. Cell. 2019;177(1):45–57.

54. Norton ME, Wapner RJ. Cell-free DNA analysis for noninvasive examination of trisomy. N Engl J Med. 2015 24;373(26):2582.

55. Johnson SB, Slade I, Giubilini A, Graham M. Rethinking the ethical principles of genomic medicine services. Eur J Hum Genet EJHG. 2020;28(2):147–54.

Library Construction for NGS

3

Melanie Kappelmann-Fenzl

Contents

What You Will Learn in This Chapter

After finishing this chapter, you will have a basic overview of the individual experimental work steps regarding the NGS library preparation workflow, the connection between the scientific or clinical question and the choice of the corresponding library preparation. For this purpose, the technical and molecular biological relevance of the individual work steps is briefly described.

M. Kappelmann-Fenzl (✉)
Deggendorf Institute of Technology, Deggendorf, Germany

Institute of Biochemistry (Emil-Fischer Center), Friedrich–Alexander University Erlangen–Nürnberg, Erlangen, Germany
e-mail: melanie.kappelmann-fenzl@th-deg.de

© Springer Nature Switzerland AG 2021
M. Kappelmann-Fenzl (ed.), *Next Generation Sequencing and Data Analysis*, Learning Materials in Biosciences, https://doi.org/10.1007/978-3-030-62490-3_3

3.1 Introduction

Library preparation involves generating a collection of DNA/cDNA fragments for sequencing. NGS libraries are typically prepared by fragmenting a DNA or RNA sample and ligating specialized adapters to both fragments ends. In this textbook we will focus on the Illumina® Library Preparation workflow for short-read sequencing (https://emea. illumina.com/techniques/sequencing/ngs-library-prep.html). As already mentioned, a wide variety of NGS methods exists, and for almost every method a comprehensive sequencing library preparation solution. The principle is almost similar for all library preparation workflows [1, 2]. Further information on library preparation for long-read [3] or single-cell sequencing [4–6] can be found on the websites of the corresponding providers:

Single-cell sequencing

- Illumina (https://emea.illumina.com/techniques/sequencing/rna-sequencing/ultra-low-input-single-cell-rna-seq.html)
- 10xGenomics (https://support.10xgenomics.com/single-cell-gene-expression/automated-library-prep)
- Qiagen (https://www.qiagen.com/de/products/discovery-and-translational-research/next-generation-sequencing/library-preparation/qiaseq-fx-single-cell-dna-library-kit/#orderinginformation)
- And many others

Long-read sequencing:

- Illumina (https://emea.illumina.com/science/technology/next-generation-sequencing/long-read-sequencing.html)
- PacificBioscience (https://www.pacb.com/products-and-services/consumables/template-preparation-multiplexing-kits/)
- Oxford Nanopore (https://nanoporetech.com/products/kits)
- And many others

3.2 Library Preparation Workflow

The core steps in preparing RNA or DNA for NGS are:

- Fragmenting and/or sizing the target sequences to a desired length.
- Converting target to double-stranded DNA (in terms of RNA-Seq).
- Attaching oligonucleotide adapters to the ends of target fragments.
- Quantifying the final library product for sequencing.

The preparation of a high-quality sequencing library plays an important role in Next-Generation Sequencing (NGS). The first major step in preparing nucleic acids for NGS is fragmentation. The most common and effective *fragmentation methods* can be subdivided into three classes:

1. Physical fragmentation (are acoustic shearing, sonication, and hydrodynamic shear).
2. Enzymatic fragmentation (DNase I or other restriction endonuclease, non-specific nuclease, Transposase).
3. Chemical fragmentation (heat and divalent metal cation). This method is used to break up long RNA fragments, whereas the length of your RNA can be adjusted by modulating the incubation time.

But also, a PCR amplification of genetic loci of interest can be chosen. Each NGS approach has its own specific protocol. Available NGS sample preparation kits are: Illumina, New England BioLabs, KAPA Biosystems, Swift Bioscience, Enzymatics, BIOO, etc. The principle workflow after fragmentation can be briefly described by the following working steps, which are also illustrated in a simplified way in Fig. 3.1 (the numbering of each working step is analog to the numbering in Fig. 3.1):

1. *Quantification and profile your isolated DNA or RNA samples.* This is one of the most important steps after sample preparation. For sequencing the samples have to be of a very good quality, the concentration must be determined, and the fragmentation efficiency must be checked (target size for short-read sequencing is commonly 200bp–800bp) before you can go any further.
2. *Perform End Repair and size selection via AMPure XP Beads* (Beckman Coulter Genomics [7]). This process converts the overhangs resulting from fragmentation into blunt ends and serves for size selection. End Repair is not performed during RNA-Seq library preparation.
3. *In Terms of RNA-Seq Library Preparation* [8].
   Depletion of rRNA and fragmentation or polyA-capture or another procedure for isolating the RNA type of interest (depending on your research question and library preparation kit you use: https://emea.illumina.com/products/by-type/sequencing-kits/library-prep-kits.html) is necessary.
4. The RNA fragments obtained must then be transcribed into cDNA for sequencing by synthesis of the first cDNA strand followed by synthesis of the second cDNA strand.
5. *Adenylate 3'-Ends of DNA/cDNA.*
   A single "A" nucleotide is added to the 3' ends of the blunt fragments to prevent them from ligating to one another during the adapter ligation reaction. A corresponding single "T" nucleotide on the 3' end of the adapter provides a complementary overhang for ligating the adapter to the fragment.
6. *Adapter Ligation and Size Selection via AMPure XP Beads of DNA/cDNA*
   Adapter ligation is a crucial step within the NGS library preparation. Adapters are design with a single "T" nucleotide on the 3' end to recognize the "A" nucleotide

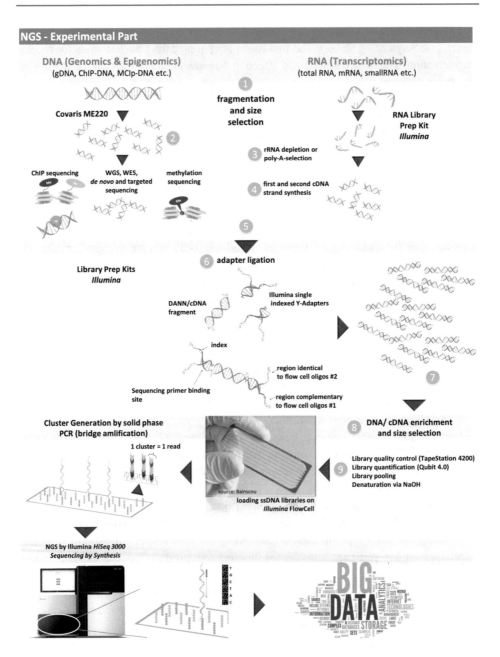

Fig. 3.1 NGS Library preparation workflow for DNA or RNA samples, respectively (source: © Melanie Kappelmann-Fenzl)

overhang (see Step 5) of each DNA/cDNA fragment. One part of the adapter sequence is complementary to the oligos covering the sequencing flow-cell and thus guarantees binding and another part is complementary to the later added sequencing primer and thus guarantees sequencing of the fragments. You can also ligate multiple indexing adapters allowing to load more than one sample onto a flow-cell. This short index enables you to distinguish between all the loaded samples carrying different indexing adapters. Adapters have a defined length of ~60 bp, hence ~120 bp long fragments can easily be identified as adapter dimers without any DNA/cDNA insert.

7. *Purify Ligation Products* (e.g., Pippin™ size selection; https://sagescience.com/products/pippin-prep/).

This process purifies the products of the ligation reaction on a gel and removes unligated adapters, as well as any adapters that might have ligated to one another.

8. *Enrich DNA/cDNA Fragments and size selection via AMPure XP Beads.*

This process uses PCR to selectively enrich those DNA/cDNA fragments that have adapter molecules on both ends and to amplify the amount of DNA/cDNA in the library. Additionally, it serves for size selection.

9. *Validate Library and normalize and pool libraries.*

This procedure is performed for quality control analysis on your sample library and for quantification. Therefore, a Agilent Technologies Bioanalyzer or TapeStation (https://www.agilent.com/en/product/automated-electrophoresis) and a Qubit 4 Fluorometer (https://www.thermofisher.com/de/de/home/industrial/spectroscopy-elemental-isotope-analysis/molecular-spectroscopy/fluorometers/qubit/qubit-fluorometer.html) are used.

This process describes how to prepare DNA/cDNA templates for cluster generation. Indexed DNA/cDNA libraries are normalized to 10nM, and then pooled in equal volumes.

Example library preparation protocols for RNA-Seq or ChIP-Seq, respectively, can be found in the Appendix section (Sect. 13.1).

Take Home Message
- Different research or clinical questions require different library preparation workflows.
- DNA sequencing approaches require proper fragmentation before library preparation.
- RNA molecules are not directly sequenced due to their chemical instability and the difficulty of processing and amplifying single-stranded nucleic acids. Thus, RNA has to be converted into cDNA (reverse transcription) for sequencing purpose.
- Quality and quantity determinations are essential within all library preparation workflows.

Review Questions

Review Question 1

Why should a final library have a median insert size of ~250–300 bp to support long paired end 2 × 150 read lengths?

Review Question 2

Which different read-outs can be obtained by different RNA-Seq Library Preparation methods?

Review Question 3

A crucial factor leading to misrepresentation of data is the bias prevalent in almost all steps of NGS sample preparation. Discuss a few possible solutions to this kind of bias!

Review Question 4

Graphically illustrate the structure of adapter-lying DNA/cDNA fragments and label the individual sequence sections of the adapters!

Answers to Review Questions

Answer to Question 1: Otherwise the percentage of adapter contaminated reads increases;

Answer to Question 2:

Objective	Principles of approach
Gene expression	Target poly(A) mRNAs (enrich or selectively amplify)
Alternative splicing	Target exon/intron boundaries by either long-read sequencing (>300 bp) or paired end sequencing (≥2 × 75) and rRNA depletion
miRNA (or small RNAs)	Target short reads (miRNAs: 18–23 bp) using size selection purification. piRNAs, snoRNAs, tRNAs are all <100 bps

Answer to Question 3: Bias during amplification of AT- and GC-rich regions: PCR-free amplification could yield better read distribution and coverage compared to PCR methods, but would require large quantities of starting DNA material.

PCR bias during library preparation for RNA-Seq can be introduced by the additional steps to convert RNA to cDNA. KAPA HiFi DNA polymerase can be used for the amplification step to reduce this kind of bias.

Answer to Question 4:

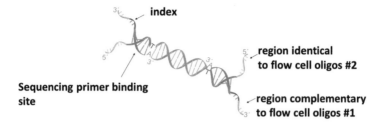

Acknowledgements We are grateful to Dr. Ines Böhme (Institute of Biochemistry (Emil-Fischer Center), Friedrich–Alexander University Erlangen–Nürnberg, Erlangen, Germany) for reviewing this chapter and Alexander Oliver Matthies (Institute of Biochemistry (Emil-Fischer Center), Friedrich–Alexander University Erlangen–Nürnberg, Erlangen, Germany) for critically reading this text.

References

1. van Dijk EL, Jaszczyszyn Y, Thermes C. Library preparation methods for next-generation sequencing: tone down the bias. Exp Cell Res. 2014;322(1):12–20.
2. Head SR, Komori HK, LaMere SA, Whisenant T, Van Nieuwerburgh F, Salomon DR, et al. Library construction for next-generation sequencing: overviews and challenges. Biotechniques. 2014;56(2):61–4, 6, 8, passim.
3. Sakamoto Y, Sereewattanawoot S, Suzuki A. A new era of long-read sequencing for cancer genomics. J Hum Genet. 2020;65(1):3–10.
4. Trombetta JJ, Gennert D, Lu D, Satija R, Shalek AK, Regev A. Preparation of Single-Cell RNA-Seq Libraries for Next Generation Sequencing. Curr Protoc Mol Biol. 2014;107:4–22. 1-17.
5. Jemt A, Salmen F, Lundmark A, Mollbrink A, Fernandez Navarro J, Stahl PL, et al. An automated approach to prepare tissue-derived spatially barcoded RNA-sequencing libraries. Sci Rep. 2016;6:37137.
6. Jemt A, Salmen F, Lundmark A, Mollbrink A, Navarro JF, Stahl PL, et al. Corrigendum: an automated approach to prepare tissue-derived spatially barcoded RNA-sequencing libraries. Sci Rep. 2017;7:41109.
7. Greenwald WW, Li H, Benaglio P, Jakubosky D, Matsui H, Schmitt A, et al. Subtle changes in chromatin loop contact propensity are associated with differential gene regulation and expression. Nat Commun. 2019;10(1):1054.
8. Podnar J, Deiderick H, Huerta G, Hunicke-Smith S. Next-Generation Sequencing RNA-Seq Library Construction. Curr Protoc Mol Biol. 2014;106:4–21. 1-19.

NGS Technologies

4

Marius Eisele and Melanie Kappelmann-Fenzl

Contents

What You Will Learn in This Chapter
There are a number of different companies that have developed and improved the NGS technology immensely in the last 10 years. In this chapter an overview of the most common technologies and their basic properties shall be given. Furthermore, it will be shown which technologies can be used for specific scientific or clinical questions and how they differ in their chemistry and output.

M. Eisele
Deggendorf Institute of Technology, Deggendorf, Germany

M. Kappelmann-Fenzl (✉)
Deggendorf Institute of Technology, Deggendorf, Germany

Institute of Biochemistry (Emil-Fischer Center), Friedrich–Alexander University Erlangen–Nürnberg, Erlangen, Germany
e-mail: melanie.kappelmann-fenzl@th-deg.de

© Springer Nature Switzerland AG 2021
M. Kappelmann-Fenzl (ed.), *Next Generation Sequencing and Data Analysis*, Learning Materials in Biosciences, https://doi.org/10.1007/978-3-030-62490-3_4

4.1   Introduction

Since the completion of the human genome project in 2003, amazing progress has been made in sequencing technologies [1]. The cost per megabase decreased and the number and diversity of sequenced genomes increased dramatically. Some approaches maximize the number of bases sequenced in the least amount of time (short-read sequencing), generating big data enabling a better understanding of complex phenotypes and disease. Alternatively, other approaches now aim to sequence longer contiguous pieces of DNA (long-read sequencing), which are essential for resolving structurally complex regions. These and other strategies are providing researchers and clinicians a variety of tools to investigate genomes, exomes, transcriptomes, epigenomes in greater depth, leading to an enhanced understanding of how biological sequence variants lead to phenotypic alterations and thus the development of various disease patterns [2].

The yearly updates of the Travis Glenn's Field Guide to Next Generation DNA Sequencer [3] are a good summary of the state of instrumentation (http://www.molecularecologist.com/next-gen-fieldguide-2016/).

4.2   Illumina

The Illumina sequencing technologies support a wide range of genetic analysis research applications, such as:

- *Whole-Genome Sequencing*: A comprehensive method for analyzing entire genomes.
- *Genotyping*: Studying variation in genetic sequences.
- *Gene Expression and Transcriptome Profiling*: Analyzing which genes and transcripts are expressed in a given sample.
- *Epigenetics*: Studying heritable changes in gene regulation that occur without a change in the DNA sequence.

Therefore, Illumina developed the Sequencing by Synthesis (SBS) Technology and BeadArray Microarray Technology. In this textbook we will focus on SBS.

The NGS massively parallel sequencing technology has revolutionized the biological sciences. With its ultra-high throughput, scalability, and speed, NGS enables researchers to perform a wide variety of applications and study biological systems at a level never before possible.

Today's complex genomic research questions demand a depth of information beyond the capacity of traditional DNA sequencing technologies. NGS has filled that gap and becomes an everyday research tool to address these questions [4]. Illumina NGS workflows include the following basic steps:

- Library Preparation (see Chap. 3)

  Libraries for NGS applications can be generated for diverse methods. Which library preparation workflow to choose depends on your scientific or clinical question and its relation to the genome, transcriptome, or epigenome of any organism. An overview of the different Illumina Library Preparation Kits can be found at https://www.illumina.com/products/by-type/sequencing-kits/library-prep-kits.html (see Chap. 3).

- Cluster Generation

  Sequencing templates are immobilized on a flow cell surface designed to present the DNA in a manner that facilitates access to enzymes while ensuring high stability of surface-bound template and low non-specific binding of fluorescently labeled nucleotides. Solid-phase amplification creates up to 1000 identical copies of each single template molecule in close proximity.

- Sequencing

  Illumina sequencing technology is also known as sequencing by synthesis (SBS) technology. Four fluorescently labeled nucleotides are used to sequence the tens of millions of clusters on the flow cell surface in parallel. During each sequencing cycle, a single labeled deoxynucleoside triphosphate (dNTP) is added to the nucleic acid chain and the nucleotide label serves as a reversible terminator for polymerization. After removing the fluorescence label of previously attached dNTP another labeled dNTP is added during a new sequencing cycle. Base calls are made directly from signal intensity measurements during each cycle.

- Data Analysis

  The NextSeq 550/2000, NextSeq 2000, and NovaSeq 6000 Sequencing Systems generate raw data files in binary base call (BCL) format, requiring conversion to FASTQ format for use with user-developed or third-party data analysis tools. Illumina offers bcl2fastq Conversion Software to convert BCL files. bcl2fastq is an included, standalone conversion software that demultiplexes data and converts BCL files to standard FASTQ files, which are the starting format for data analysis.

As already described in Chap. 3, the library, which was prepared by random fragmentation of the DNA or cDNA (in terms of RNA-Seq) sample, followed by 5′ and 3′ adapter ligation, PCR amplification, and gel purification (see Fig. 4.1a and Chap. 3). For cluster generation, the library is loaded onto a flow cell where fragments are captured on a lawn of surface-bound oligos complementary to the library adapters. Each fragment is then amplified into distinct, clonal clusters through bridge amplification. When cluster generation is complete, the templates are ready for sequencing (Fig. 4.1b). The sequencing by synthesis (SBS) technology uses a proprietary reversible terminator-based method that

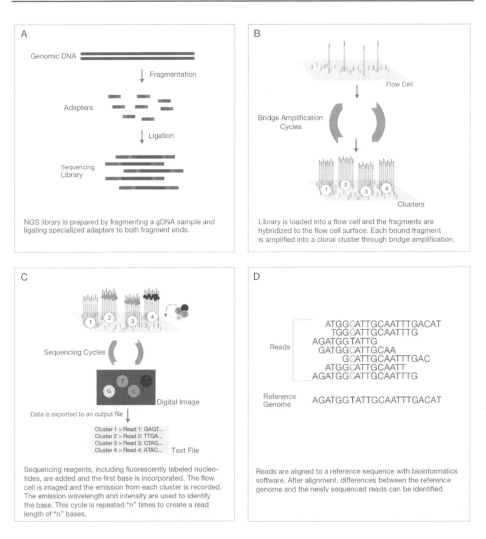

Fig. 4.1 Next Generation Sequencing Chemistry Overview—Illumina NGS includes four steps: (**a**) library preparation, (**b**) cluster generation, (**c**) sequencing, and (**d**) alignment and data analysis. (source: www.illumina.com)

detects single bases as they are incorporated into DNA template strands. As all four reversible terminator-bound dNTPs are present as single, separate molecules during each sequencing cycle, natural competition minimizes incorporation bias and greatly reduces raw error rates. The result is highly accurate base-by-base sequencing that virtually eliminates sequence context-specific errors, even within repetitive sequence regions and homopolymers (see Fig. 4.1c). During data analysis and alignment, the newly identified sequence reads are aligned to a reference genome. Following alignment, many variations of analysis are possible, such as single nucleotide polymorphism (SNP) or insertion-deletion (indel) identification, read counting for RNA methods, phylogenetic or metagenomic analysis, and more. A graphical overview of the NGS chemistry is depicted in Fig. 4.1.

4.3   Ion Torrent

Unlike Illumina Ion Torrent semiconductor sequencing from Thermo Fisher Scientific does not make use of optical signals. Instead, they exploit the fact that addition of a dNTP to a DNA polymer releases an H^+ ion.

As in other kinds of NGS, the input DNA or RNA is fragmented to approximately 200bp, adapters are added, and one molecule is placed onto a bead. The molecules are amplified on the bead by emulsion PCR resulting in millions of different beads with millions of different fragments. These beads than flow across the semiconductor chip depositing each bead into a single well. Next the slide is flooded with a single species of dNTP (one NTP at a time), along with buffers and polymerase. The pH is detected, as each H^+ ion released will decrease the pH. The changes in pH allow to determine if that base, and how many thereof, was added to the sequence read. The dNTPs are washed away and the process is repeated cycling through the different dNTP species. The pH change (if any) is utilized to determine how many bases (if any) were added with each cycle.

If a nucleotide, for example, a C, is added to a DNA template and is then incorporated into a strand of DNA, a hydrogen ion will be released. The charge from that ion will change the pH of the solution in the well, which can be detected by a specific ion sensor. This process happens simultaneously in millions of wells, that is why this technology is often described as massively parallel sequencing.

The Ion Torrent NGS instruments Genexus, Ion GeneStudio S5, ION PGM Dx, Ion Chef, and Ion OneTouch2 are essentially the world's smallest solid-state pH meters, calling the base, going directly from chemical information to digital information.

4.4   Pacific Bioscience

Single-molecule, real-time (SMRT) sequencing developed by Pacific BioSciences (PacBio) offers longer read lengths than the second-generation sequencing (SGS) technologies, making it well-suited for unsolved problems in genome, transcriptome, and epigenetics research [5].

Introducing the PacBio Sequel II system powered by SMRT sequencing technology the first step is to isolate DNA or RNA from any sample type. Next a SMRTbell library is created by ligating hairpin adapters to double stranded DNA creating a circular template. Primer and polymerase are added to the library that is placed on the instrument for sequencing. The smart cell contains millions of small, tiny wells called zero-mode waveguides (ZMWs). A single molecule of DNA is immobilized in a ZMW sequencing

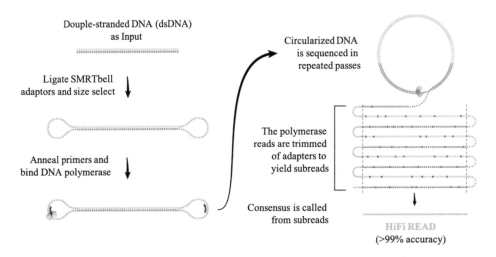

Fig. 4.2 Using the circular consensus sequencing (CCS) mode for HiFi READ production to provide base-level resolution with >99% single-molecule read accuracy for the detection of all variant types from single nucleotide to structural variants (source: modified according to https://www.pacb.com)

unit, which provides the smallest available volume for light detection and as the polymerase incorporates fluorescently labeled deoxyribonucleoside triphosphates (dNTPs) light is emitted. The order of their enzymatic incorporation into a growing DNA strand is detected via ZMW nanostructure arrays, which allow the simultaneous detection of thousands of single-molecule sequencing reactions. The replication processes in all ZMWs of a SMRT cell are recorded by a "movie" of light pulses, and the pulses corresponding to each ZMW can be interpreted to be a sequence of bases. With this approach nucleotide incorporation is measured in real time. With the Sequel II system you can optimize your results with two sequencing modes. You can use the circular consensus sequencing (CCS) mode to produce highly accurate long reads, known as HiFi reads (Fig. 4.2), or use the continuous long-read sequencing (CLR) mode to generate the longest possible reads (Fig. 4.3). The average read length from the PacBio instrument is approximately 2 kb, and some reads may be over 20 kb. Longer reads are especially useful for de novo assemblies of novel genomes as they can span many more repeats and bases.

4.5   Oxford Nanopore

In essence, Oxford Nanopore is a real-time, high-throughput technology and is specialized on long-read and single-molecule sequencing. Oxford Nanopore technology consists of millions of nanoscale pores spanned across an impermeable thin membrane, allowing massively parallel sequencing. The membrane separates two chambers, both contain an electrolyte and a single connection to each other via a single nanopore. The applied voltage

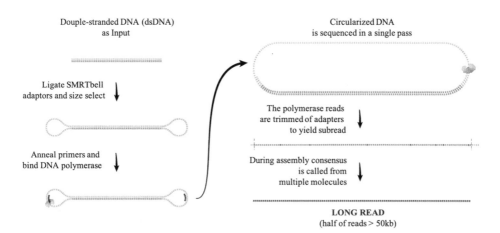

**Double-stranded DNA (dsDNA)
as Input**

Ligate SMRTbell
adaptors and size select

Anneal primers and
bind DNA polymerase

**Circularized DNA
is sequenced in a single pass**

The polymerase reads
are trimmed of adapters
to yield subread

During assembly consensus
is called from
multiple molecules

LONG READ
(half of reads > 50kb)

Fig. 4.3 Using the continuous long-read (CLR) sequencing mode for sequence read lengths in the tens of kilobases to enable high-quality assembly of even the most complex genomes. With SMRT sequencing you can expect half the data in reads >50 kb and the longest reads up to 175 kb (source: modified according to https://www.pacb.com)

by two electrodes generates an ion flow from one chamber, through the pore and into the other chamber.

This way, ions and charged biomolecules like the nucleic acid molecules with their negative charge can be driven through the pore. The ions act as a motor, allowing the molecules to be passed through the channel. Consequently, structural features, such as the bases or the epigenetic modification of the sequences, can be identified by tracing the ionic current, which is partially blocked by the molecule. Compared to other sequencing technologies, Oxford Nanopore with its fascinating simple biophysical approach has resulted in overwhelming academic, industrial, and national interest (Fig. 4.4, [6]).

Historically, the pioneer technology giving rise to the Oxford Nanopore was invented by Wallace H. Coulter in the late 1940s. Coulter's technology was using essentially the same basic chemo-physical principle as Oxford Nanopore, but was used for counting and sizing blood cells. An automated version of Coulter's counters is still used in hospitals today. However, the true reincarnation of the Coulter's counters was in the 1990s, when the pore was not of millimeter but of nanometer dimensions, allowing the analysis of ions and biomolecules instead of whole cells [7].

Properties that an analyte should have: Every analyte molecule consists of multiple ions that allow it to pass the pore. Furthermore, the pore has to be wide enough (i.e., around 2 nm) and must permit the transport of ions. Ultimately, the flow of ions across the pore should be able to report on subtle differences between the analytes.

There are two types of pores that are currently being used: protein and solid-state channels.

Examples of protein channels are toxin α-hemolysin, which is secreted by *Staphylococcus aureus*, and MspA from *Mycobacterium smegmatis*. A promising approach for solid-state channels is the use of TEM (Transmission electron microscopy), combined with a

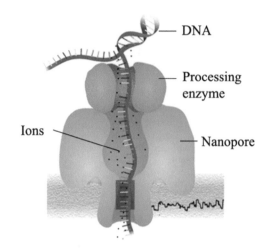

Fig. 4.4 Graphic representation of DNA sequencing using a MinION. A processive enzyme (green) ratchets DNA into the pore (blue), causing a change in ionic current (ions are shown as black dots) that is determined by the 6-mer in the central channel (purple box). The current is recorded over time (black trace, bottom right). (modified according to Muller et al. [6])

single-layer graphene membrane, however, up to this point (2020) the protein channels are superior to the solid-state channels [8].

As mentioned earlier, biomolecules that pass through the pore generate the signal by partially blocking the flow of ions, which can then be translated into the sequence and epigenetic modifications. Nevertheless, ions lining up at the membrane, together with the counterions on the opposite site of the membrane, also contribute to the signal, generating noise. These noise fluctuations increase with bandwidths, which limits the time resolution in experiments. Apart from shorter measurement times, a common way to compensate the noise is achieved by using analog or digital low-pass filters. Still, the generation of noise, introducing error-prone data, might be the biggest struggle Oxford Nanopore Technology has yet to overcome.

Big advantages of this sequencing technology compared to others on the market are its portability and price tag. Oxford Nanopores' smallest device, the MinION is controlled and powered by an USB cable and is just slightly bigger than a regular USB stick. Depending on the experiment (DNA or RNA), Oxford Nanopore devices do not need an amplification step (PCR) prior to the sequencing. Theoretically, the only limitation in sequencing length is the time and therefore the induced noise. So far, the maximum of usable read length is around 100 kilobases [9]. However, longer reads result in less accurate data [10].

4.6 NGS Technologies: An Overview

A more detailed overview of the major NGS platforms and their general properties [10] are listed in the Table in the Appendix section (Table 13.1: Major NGS platforms and their general properties.)

Take Home Message

- The term NGS is used to summarize second- and third-generation sequencing methods.
- Basically, a distinction can be made between first-generation sequencing (Sanger sequencing), second-generation sequencing (massively parallel sequencing), and third-generation sequencing (single-molecule sequencing).
- With regard to sequencing technologies, one also differentiates between short- and long-read sequencing.
- The different NGS technologies entail different preprocessing and analysis steps and are applied depending on the scientific or clinical question to the resulting data.
- Each NGS technology can be characterized by its input template, read length, error rate, sequencing scheme, visualization method, sequencing principle, and amount of data output.

Further Reading

- See Table 4.1: Sequencing technologies of some companies and their products.

Table 4.1 Sequencing technologies of some companies and their products

	Products	Technology	Further Information
Illumina	iSeq 100, MiniSeq, MiSeq, NextSeq 550, NextSeq 2000, NovaSeq 6000	SBS, short-read	https://emea. illumina.com/
10xGenomics	Chromium Controller	Linked reads	https://www. 10xgenomics. com/
Thermo Fisher	Ion GeneStudio S5 series Ion PGM Dx System	SBS, short-read	https://www. thermofisher. com/
Pacific Bioscience	PacBio Sequel Systems	Long-read sequencing (Up to 160Gb), Single-Molecule, Real-Time (SMRT) sequencing	https://www. pacb.com/
Qiagen	GeneReaderPlatform	End-to-end NGS workflow, clinical testing	https://www. qiagen.com/
Oxford Nanopore	MinION (Up to 30Gb) GridION X5 (Up to 150Gb) PromethION (Up to 9,600 Gb)	Long-read (Up to 2Mb), real-time nanopore sequencing	https:// nanoporetech. com/

Review Questions

1. Which of the following statements regarding the quantity of template for a sequencing reaction is correct?
   A. Excess template reduces the length of a read.
   B. Too little template will result in very little readable sequences.
   C. Excess template reduces the quality of a read.
   D. All of the above.
2. What will heterozygous single nucleotide substitution look like on your chromatogram?
   A. Two peaks of equal height at the same position.
   B. One peak twice the height of those around it.
   C. Two peaks in the same position, one twice the height of the other.
   D. Three peaks of equal height at the same position.
3. Which of the following is important for preparing templates for Next Generation Sequencing?
   A. Isolating DNA from tissue.
   B. Breaking DNA up into smaller fragments.
   C. Checking the quality and quantity of the fragment library.
   D. All of the above.
4. Which of the below sequencing techniques require DNA amplification during the library preparation step (is considered a 2nd generation sequencing technique)?
   A. PacBio AND Oxford Nanopore.
   B. Illumina AND Ion Torrent.
   C. Illumina AND Oxford Nanopore.
   D. PacBio AND Ion Torrent.
5. Which of the below sequencing techniques use(s) fluorescently labeled nucleotides for identifying the nucleotide sequence of the template DNA strand?
   A. Illumina AND PacBio.
   B. Illumina AND Oxford Nanopore.
   C. PacBio AND Ion Torrent.
   D. Only Illumina.
   E. All sequencing methods use fluorescently labeled nucleotides for identifying the nucleotide sequence of the template DNA strand.
6. The below figures illustrate five cycles of Illumina sequencing. The colored spots represent the clusters on the flow cell. What is the sequence of the DNA template (cluster) in the top, left corner according to the figure?
   A: Yellow
   C: Red
   G: Blue
   T: Green

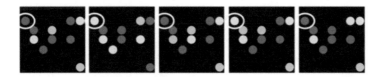

7. What is the main enzyme component of Sanger sequencing?
8. Which of the following best describes three cyclic steps in PCR, in the correct order?
   A. Denaturing DNA to make it single stranded, primer annealing, synthesis of a new DNA strand.
   B. Primer annealing, denaturing DNA to make it single stranded, synthesis of a new DNA strand.
   C. Synthesis of a new DNA strand, primer annealing, denaturing DNA to make it single stranded.
9. Which of the following "omes" relates to the DNA sequence of expressable genes?
   A. Genome.
   B. Exome.
   C. Proteome.
   D. Metabolome.
10. Targeted sequencing:
    A. Allows to focus on specific areas of interest and thus enables sequencing at much higher coverage levels.
    B. Is a sequencing method used within molecular diagnostics.
    C. Is equivalent to sequencing tumor panels.
    D. This method refers to sequence a novel genome.
11. Library preparation involves generating a collection of DNA fragments for sequencing. NGS libraries are typically prepared by fragmenting a DNA or RNA sample and ligating specialized adapters to both fragments ends. Bring the following working steps in terms of library preparation in the right order:
    A. Perform End Repair and size selection via AMPure XP Beads.
    B. Quantification and profile samples.
    C. Purify Ligation Products.
    D. Validate Library.
    E. Adenylate 3'-Ends.
    F. Normalize and pool libraries.
    G. Enrich DNA Fragments and size selection via AMPure XP Beads.
    H. Adapter Ligation and size selection via AMPure XP Beads.
12. Indicate whether each of the following descriptions better applies to Illumina® sequencing (I), Ion Torrent™ sequencing (T), or both sequencing technologies (B).
    – It uses fluorescently labeled nucleotides.
    – It uses PCR-generated copies of DNA.

- It is a second-generation sequencing technology and employs a cyclic wash-and-measure paradigm.
- It relies on the fidelity of DNA polymerase for its accuracy.
- pH change indicates how many bases were added.

Answers to Review Questions

1D; 2A; 3D; 4B; 5A; 6 GAGAC; 7 Polymerase; 8A; 9B; 10A, B, C; 11 BAEHCGDF; 12 IBBBT;

Acknowledgements We are grateful to Melinda Hutcheon, (Illumina, Inc., 5200 Illumina Way, San Diego, CA 92122, US) for reviewing this chapter. We would also like to thank *Illumina* for the material and illustrations provided to help us write this chapter.

References

1. Hood L, Rowen L. The human genome project: big science transforms biology and medicine. Genome Med. 2013;5(9):79.
2. Goodwin S, McPherson JD, McCombie WR. Coming of age: ten years of next-generation sequencing technologies. Nat Rev Genetics. 2016;17(6):333–51.
3. Glenn TC. Field guide to next-generation DNA sequencers. Mol Ecol Resour. 2011;11 (5):759–69.
4. Morganti S, Tarantino P, Ferraro E, D'Amico P, Viale G, Trapani D, et al. Complexity of genome sequencing and reporting: next generation sequencing (NGS) technologies and implementation of precision medicine in real life. Crit Rev Oncol Hematol. 2019;133:171–82.
5. Rhoads A, Au KF. PacBio sequencing and its applications. Genomics Proteomics Bioinformatics. 2015;13(5):278–89.
6. Muller CA, Boemo MA, Spingardi P, Kessler BM, Kriaucionis S, Simpson JT, et al. Capturing the dynamics of genome replication on individual ultra-long nanopore sequence reads. Nat Methods. 2019;16(5):429–36.
7. Wanunu M. Nanopores: a journey towards DNA sequencing. Phys Life Rev. 2012;9(2):125–58.
8. Jain M, Olsen HE, Paten B, Akeson M. The Oxford nanopore MinION: delivery of nanopore sequencing to the genomics community. Genome Biol. 2016;17(1):239.
9. De Coster W, Strazisar M, De Rijk P. Critical length in long-read resequencing. NAR Genomics Bioinform. 2020;2(1):lqz027.
10. McCombie WR, McPherson JD, Mardis ER. Next-generation sequencing technologies. Cold Spring Harb Perspect Med. 2019;9(11):a033027.

Computer Setup

5

Melanie Kappelmann-Fenzl

Contents

What You Will Learn in This Chapter
This chapter describes the minimum hardware and software requirements to analyze Next-Generation Sequencing data. There are various solutions to set up a compatible computer environment. Here, the basic installation steps are clearly presented and the fields of application of some important bioinformatic tools are described.

M. Kappelmann-Fenzl (✉)
Deggendorf Institute of Technology, Deggendorf, Germany

Institute of Biochemistry (Emil-Fischer Center), Friedrich-Alexander University Erlangen-Nürnberg, Erlangen, Germany
e-mail: melanie.kappelmann-fenzl@th-deg.de

© Springer Nature Switzerland AG 2021
M. Kappelmann-Fenzl (ed.), *Next Generation Sequencing and Data Analysis*, Learning Materials in Biosciences, https://doi.org/10.1007/978-3-030-62490-3_5

5.1   Introduction

Many NGS data analysis programs are written in Java, Perl, C++ or Python. Simply running these programs does not require programming knowledge, however, it is indeed helpful. Within the framework of this textbook bioinformatic details are not covered. Anyway, to be able to perform NGS data analysis familiarity with R is important. For those of you who are not familiar with R, it is highly recommended to take advantage of some freely available web resources (e.g. https://www.bioconductor.org/help/course-materials/2012/SeattleMay2012/).

5.2   Computer Setup for NGS Data Analysis

Next-generation sequencing analysis is a computationally demanding process. Your average laptop is probably not up to the challenge. Any Linux/Unix-based operating system will work well, with large servers in mind. Typically, analysis algorithms will be distributed by researchers in one of the three ways:

- Standalone program to run in a Linux/Unix environment (most common).
- Webserver where you upload data to be analyzed (i.e. Galaxy).
- R software package/library for the R computing environment.

In theory, any computer with enough RAM, hard drive space, and CPU power can be used for analysis. In general, you will need:

- 16 Gb of RAM Minimum (better to have 96+ Gb).
- 500 Gb of disk space (better to have 10+ Tb of space).
- Fast CPU (better to have at least 8 cores, more the better).
- External storage.

Depending on the type of analysis you want to perform, the required RAM, hard drive space, and CPU power may vary and for some analyses a conventional laptop is fully sufficient. If you own a computer or lab server with the aforementioned properties, you can start over setting up your computer environment. The command line analysis tools that we demonstrate in this book run on the Linux/Unix operating system. For practicing NGS data analysis, we will use several software tools in various formats:

- Binary *executable* code that can be run directly on the target computer.
- *Source code* that needs to be compiled to create a *binary* program.
- Programs that require the presence of another programming language like Java, Python, or Perl.

Setting up your computer takes a little while and can be a bit tedious, but you only need to do this once.

5.3   Installation

If you are working with MacOS, you need to change the *shell* to *Bash* and install XCode (App Store) with the *Command Line tools* package which includes GNU compilers and many other standard Linux/Unix development tools.

```
#check which shell you are currently using
echo "$SHELL"
#change shell to Bash
chsh -s /bin/bash
#check if shell changed to Bash
echo "$SHELL"
#after installing XCode install the additional command line tools to XCode
xcode-select --install
```

Next, you have to initialize the terminal.

```
#create .bash_profile and .bashrc
cd ~
> .bash_profile
> .bashrc
ls -a
```

Set up your .bash_profile, by adding the following to the *.bash_profile* file and *.bashrc* file.
.bash_profile:

```
cd ~
cat >> .bash_profile << EOF
if [ -f ~/.bashrc ]; then
source ~/.bashrc
fi
EOF
#check if it worked
cat .bash_profile
```

.bashrc file:

```
cd ~
cat >> .bashrc << EOF
#A minimal BASH profile.
#Will ask permissions before overwriting files.
alias rm='rm -i'
alias mv='mv -i'
alias cp='cp -i'
#Extend the program search PATH and add the ~/bin folder.
export PATH=~/bin:$PATH
#You can make the prompt more user friendly.
export PS1='\[\e]0;\w\a\]\n\[\e[32m\]\u@\h \[\e[33m\]\w\[\e[0m\]\n\$ '
#This is required for Entrez Direct to work.
export PERL_LWP_SSL_VERIFY_HOSTNAME=0
EOF
#check if it worked
cat .bashrc
#apply new settings in .bash_profile to your current terminal
source ~/.bash_profile
```

We will set up a computer environment using *Bioconda* (a channel for the Conda package manager specializing in bioinformatics software; https://bioconda.github.io/).

Hint: Installation steps via command line tool are finished by reopening a new terminal window!

Open a Terminal and execute the following commands:

```
#Download installer
#For MacOS
curl -OL https://repo.anaconda.com/miniconda/Miniconda3-latest-MacOSX-
x86_64.sh
#For Linux
curl -OL https://repo.anaconda.com/miniconda/Miniconda3-latest-Linux-
x86_64.sh
#run installer
#For MacOS
bash Miniconda3-latest-MacOSX-x86_64.sh
#For Linux
bash Miniconda3-latest-Linux-x86_64.sh
#Scroll down Licence agreement and type "yes" if asked
#Check version
conda -V
#update conda package to bring it to the latest version:
conda update -y -n base conda
```

```
#check if version changed
conda -V
#close and reopen terminal to activate latest conda version
#activate the bioconda channel within conda
conda config –add channels bioconda
conda config –add channels conda-forge
```

It is also possible to create a "bioinformatic environment" within *Bioconda*, where all software tools used for your analysis are "stored," but you do not have to! For more details read: https://uoa-eresearch.github.io/eresearch-cookbook/recipe/2014/11/20/conda/.

Due to the fact that we are using our computer only for bioinformatic purposes, we just add all software tools into the *bin* folder of *miniconda3* and add the path to our *.bashrc*.

Some other helpful commands:

```
# upgrade conda packages
conda update "toolname"
# update conda itself
conda update -n base condacon
# apply new settings in .bash_profile to your current terminal
source ~/.bash_profile
```

A list of all available *Bioconda* software tools can be found at https://bioconda.github.io/conda-recipe_index.html.

If you want to install some new tools via run `conda install "bioconda toolname"` in the terminal.

The following table (Table 5.1) contains the most important and helpful tools to perform NGS data analysis.

The most time saving and easiest way to download the bioinformatic packages via *Conda* is to create a .txt file with all the package names.

```
#install all listed packages in the .txt file via conda
cat bioconda_packages.txt | xargs conda install -y
```

Another important tool for Variant discovery based on NGS data is the Genome Analysis Toolkit. The latest download release can be found here (https://github.com/broadinstitute/gatk/releases/download/4.1.8.0/gatk-4.1.8.0.zip).

Table 5.1 Bioinformatics related packages for Linux and Mac OS provided by Bioconda

Bioconda software tool	Short description	Further information
bamtools [1]	C++ API & command-line toolkit for working with BAM data	https://github.com/pezmaster31/bamtools
bbmap	BBMap is a short-read aligner, as well as various other bioinformatic tools	https://sourceforge.net/projects/bbmap
BCFtools [2, 3]	BCFtools is a set of utilities that manipulate variant calls in the Variant Call Format (VCF)	https://github.com/samtools/bcftools
BEDTools [4, 5]	A powerful toolset for genome arithmetic	http://bedtools.readthedocs.org/
bioawk	BWK awk modified for biological data	https://github.com/lh3/bioawk
blast	BLAST+ is a new suite of BLAST tools that utilizes the NCBI C++ Toolkit	http://blast.ncbi.nlm.nih.gov
bowtie2 [6]	Fast and sensitive read alignment Mapping MCiP/ChIP-Seq data	http://bowtie-bio.sourceforge.net/bowtie2/index.shtml
bwa [7, 8]	The BWA read mapper	https://github.com/lh3/bwa
cutadapt	Trim adapters from high-throughput sequencing reads	https://cutadapt.readthedocs.io/
datamash	GNU Datamash is a command-line program which performs basic numeric, textual and statistical operations on input textual data files	http://www.gnu.org/software/datamash
Emboss [9]	The European Molecular Biology Open Software Suite	http://emboss.open-bio.org/
entrez-direct	Entrez Direct (EDirect) is an advanced method for accessing the NCBI's set of interconnected databases (publication, sequence, structure, gene, variation, expression, etc.) from a Linux/Unix terminal window	ftp://ftp.ncbi.nlm.nih.gov/entrez/entrezdirect/versions/13.3.20200128/README
fastqc	A quality control tool for high-throughput sequence data	http://www.bioinformatics.babraham.ac.uk/projects/fastqc/
freebayes	Bayesian haplotype-based polymorphism discovery and genotyping	https://github.com/ekg/freebayes
control-freec	Copy number and genotype annotation from whole genome and whole exome sequencing data	https://github.com/BoevaLab/FREEC
hisat2 [10]	Graph-based alignment of next-generation sequencing reads to a population of genomes	https://ccb.jhu.edu/software/hisat2/index.shtml
htslib	C library for high-throughput sequencing data formats	https://github.com/samtools/htslib

(continued)

Table 5.1 (continued)

Bioconda software tool	Short description	Further information
minimap2	Experimental tool to find approximate mapping positions between long sequences	https://github.com/lh3/minimap
perl	The Perl programming language interpreter	http://www.perl.org/
perl-net-ssleay	Perl extension for using OpenSSL	http://metacpan.org/pod/Net::SSLeay
picard	Java tools for working with NGS data in the BAM format	http://broadinstitute.github.io/picard/
readseq	Read & reformat biosequences, Java command-line version	http://iubio.bio.indiana.edu/soft/molbio/readseq/java/
samtools [11]	Tools for dealing with SAM, BAM and CRAM files	https://github.com/samtools/samtools
seqkit	Cross-platform and ultrafast toolkit for FASTA/Q file manipulation	https://github.com/shenwei356/seqkit
seqtk	Seqtk is a fast and lightweight tool for processing sequences in the FASTA or FASTQ format	https://github.com/lh3/seqtk
snpeff	Genetic variant annotation and effect prediction toolbox	http://snpeff.sourceforge.net/
sra-tools	Retrieve raw data from NCBI-SRA; download data files directly	https://github.com/ncbi/sra-tools
Star	Mapping RNA-Seq data. Splice aware	https://github.com/alexdobin/STAR
subread	High-performance read alignment, quantification, and mutation discovery	http://subread.sourceforge.net/
trimmomatic	A flexible read trimming tool for Illumina NGS data	http://www.usadellab.org/cms/?page=trimmomatic
vt	A tool set for manipulating and generating VCF files	https://genome.sph.umich.edu/wiki/Vt
wget	Free utility for non-interactive download of files from the Web	https://www.gnu.org/software/wget/manual/wget.html
Other essential/ helpful software tools		
R and RStudio	Statistical analysis program	https://www.rstudio.com/
Homer [12]	Basic ChIP Seq analysis: finding peaks/regions; genome annotation of peaks; functional annotation; Motif finding, etc.	http://homer.ucsd.edu/homer/
GSEA [13, 14]	Gene Set Enrichment Analysis	http://software.broadinstitute.org/gsea/index.jsp
GREAT	Function prediction of cis-regulatory regions	http://great.stanford.edu/public/html/index.php

(continued)

Table 5.1 (continued)

Bioconda software tool	Short description	Further information
GATK	Genome Analysis Toolkit for Variant Discovery in High-Throughput Sequencing Data	https://gatk.broadinstitute.org/hc/en-us
IgV [15–17]	Visualization of sequencing reads	https://software.broadinstitute.org/software/igv/
Text Editor	Supports working with tab-delimited data	

Once you have downloaded and unzipped the package (named gatk-[version]), you will find four files inside the resulting directory:

```
gatk
gatk-package-[version]-local.jar
gatk-package-[version]-spark.jar
README.md
```

Then export the PATH where you store the GATK package to *.bashrc*. `export PATH="/path/to/gatk/:$PATH"` where /path/to/gatk is the path to the location of the gatk executable. Note that the jars must remain in the same directory as gatk for it to work.

Check if it works:

```
gatk --list
```

Next, you have to install R (e.g. https://packages.othr.de/cran/) and RStudio (https://www.rstudio.com/) and set up the local R library path. If you want to use another location rather than the default location, for example, ~/local/R_libs/ you need to create the directory first:

```
mkdir ~/R_libs
```

Open R or RStudio and install the package *lattice* from the console by defining the path to the newly created R_libs folder.

```
> install.packages("lattice", repos="http://cran.r-project.org", lib="~/R_libs/")
```

It is a bit of burden having to type the long string of library path every time. To avoid doing that, you can create a file .Renviron in your home directory, and add the following content to the file:

```
cd ~
> .Renviron
cat >> .Renviron << EOF
R_LIBS=~/R_libs
EOF
```

Whenever R is started, the directory ~/R_libs/ is added to the list of places to look for R packages. To see the directories where R searches for libraries type `>.libPaths()`.

```
>.libPaths("~/R_libs")
> install.packages("lattice", repos="http://cran.r-project.org")

#will have the same effect as the previous install.packages() command.

#check the R library paths

>.libPaths()
```

After installation of R and definition of your R_libs folder you can install Bioconductor (https://www.bioconductor.org/install/).

```
if (!requireNamespace("BiocManager", quietly = TRUE))
  install.packages("BiocManager")
```

After installation of *BiocManager* you can search through all available packages by typing `BiocManager::available()`, and install one or more packages by `BiocManager::install()`.

Next, it is highly recommended to set up an organized file structure. Therefore, the usage of short and descriptive folder- and filenames should be used (see Chap. 6).

Take Home Message
- NGS data analyses are computationally intensive.
- Bioconda is a channel for the Conda package manager specializing in bioinformatics software.
- To run other NGS data analysis tools not installed via conda set the path to the executable files in your *.bashrc*.

Further Reading The Biostar Handbook: 2nd Edition. https://www.biostarhandbook.com/.

Acknowledgements We thank Patricia Basurto and Carolina Castañeda ot the International Laboratory for Human Genome Research (National Autonomous University of Mexico, Juriquilla campus) for reviewing this chapter.

References

1. Barnett DW, Garrison EK, Quinlan AR, Stromberg MP, Marth GT. BamTools: a C++ API and toolkit for analyzing and managing BAM files. Bioinformatics. 2011;27(12):1691–2.
2. Danecek P, McCarthy SA. BCFtools/csq: haplotype-aware variant consequences. Bioinformatics. 2017;33(13):2037–9.
3. Narasimhan V, Danecek P, Scally A, Xue Y, Tyler-Smith C, Durbin R. BCFtools/RoH: a hidden Markov model approach for detecting autozygosity from next-generation sequencing data. Bioinformatics. 2016;32(11):1749–51.
4. Quinlan AR. BEDTools: The Swiss-Army tool for genome feature analysis. Curr Protoc Bioinformatics. 2014;47:11 2 1–34.
5. Quinlan AR, Hall IM. BEDTools: a flexible suite of utilities for comparing genomic features. Bioinformatics. 2010;26(6):841–2.
6. Langdon WB. Performance of genetic programming optimised Bowtie2 on genome comparison and analytic testing (GCAT) benchmarks. BioData Min. 2015;8(1):1.
7. Li H, Durbin R. Fast and accurate long-read alignment with Burrows-Wheeler transform. Bioinformatics. 2010;26(5):589–95.
8. Li H, Durbin R. Fast and accurate short read alignment with Burrows-Wheeler transform. Bioinformatics. 2009;25(14):1754–60.
9. Rice P, Longden I, Bleasby A. EMBOSS: the European molecular biology open software suite. Trends Genet. 2000;16(6):276–7.
10. Kim D, Langmead B, Salzberg SL. HISAT: a fast spliced aligner with low memory requirements. Nat Methods. 2015;12(4):357–60.
11. Li H, Handsaker B, Wysoker A, Fennell T, Ruan J, Homer N, et al. The sequence alignment/map format and SAMtools. Bioinformatics. 2009;25(16):2078–9.
12. Heinz S, Benner C, Spann N, Bertolino E, Lin YC, Laslo P, et al. Simple combinations of lineage-determining transcription factors prime cis-regulatory elements required for macrophage and B cell identities. Molecular Cell. 2010;38(4):576–89.

13. Mootha VK, Lindgren CM, Eriksson KF, Subramanian A, Sihag S, Lehar J, et al. PGC-1alpha-responsive genes involved in oxidative phosphorylation are coordinately downregulated in human diabetes. Nat Genet. 2003;34(3):267–73.
14. Subramanian A, Tamayo P, Mootha VK, Mukherjee S, Ebert BL, Gillette MA, et al. Gene set enrichment analysis: a knowledge-based approach for interpreting genome-wide expression profiles. Proc Natl Acad Sci U S A. 2005;102(43):15545–50.
15. Robinson JT, Thorvaldsdottir H, Wenger AM, Zehir A, Mesirov JP. Variant review with the integrative genomics viewer. Cancer Res. 2017;77(21):e31–e4.
16. Thorvaldsdottir H, Robinson JT, Mesirov JP. Integrative Genomics Viewer (IGV): high-performance genomics data visualization and exploration. Brief Bioinform. 2013;14(2):178–92.
17. Robinson JT, Thorvaldsdottir H, Winckler W, Guttman M, Lander ES, Getz G, et al. Integrative genomics viewer. Nat Biotechnol. 2011;29(1):24–6.

Introduction to Command Line (Linux/Unix)

6

Melanie Kappelmann-Fenzl

Contents

What You Will Learn in This Chapter

This chapter will introduce you in the Linux/Unix file system and how to navigate your computer environment and file system tree from the terminal. Moreover, you will become familiar with basic Linux/Unix commands and their purpose and how to manipulate certain files.

M. Kappelmann-Fenzl (✉)
Deggendorf Institute of Technology, Deggendorf, Germany

Institute of Biochemistry (Emil-Fischer Center), Friedrich-Alexander University Erlangen-Nürnberg, Erlangen, Germany
e-mail: melanie.kappelmann-fenzl@th-deg.de

© Springer Nature Switzerland AG 2021
M. Kappelmann-Fenzl (ed.), *Next Generation Sequencing and Data Analysis*, Learning Materials in Biosciences, https://doi.org/10.1007/978-3-030-62490-3_6

6.1 Introduction

Most computer users today are familiar with the graphical user interphase (GUI), however, it has been said that "graphical user interfaces make easy tasks easy, while the command line interfaces (CLI) make difficult tasks possible." That is reason enough to spend more time with the CLI, as we often have to deal with more difficult issues within NGS data analysis.

When we speak of the command line interface, we actually are referring to the *shell*, which is a program that takes keyboard commands and passes them to the operating system to carry out. Most Linux distributions supply a *shell* program from the GNU Project called *Bash* (**b**ourne-**a**gain *shell*), which is a replacement for *sh* (the original Unix *shell* program). Bioinformatics is a lot about manipulating text files, thus the usage of the command line interpreter (CLI) or *shell* is an essential tool. CLI or *shell* is a program that parses and executes commands from the user in the terminal. A common command line interpreter is *Bash*.

6.2 The Linux/Unix File System

The purpose of this chapter is to introduce you to how files and directories are handled in Linux/Unix and how to navigate and manipulate your file system from the command line tool (terminal). If you are opening your terminal (Unix: Applications→ Utilities→ double-click Terminal; Linux: Strg + Alt + T or search in Applications/Accessories → Terminal), you are placed in your "home" directory, which is than your "present working directory" (pwd). In your home or another present working directory, you can create files and subdirectories. The commands that you issue in the terminal at the Unix prompt ($) relate to the files and folders and resources available from your present working directory. Understanding the Linux/Unix file system also enables you to use and refer to resources outside of your current working directory. The Linux/Unix file system is depicted in Fig. 6.1.

All files are organized in a tree-like structure starting at the *root* folder /.

Purpose of each of directory:

- usr/bin: Most commands and executable files.
- /dev: Device entries for disks, printers, pseudo-terminals, etc.
- /etc: Critical startup and configuration files.
- /sbin: Commands needed for minimal system operability.
- /home: Default home directories for users.
- /var: System specific data and configuration files.
- /tmp: Temporary files that may disappear between reboots.
- /opt: Optional software packages (not consistently used).

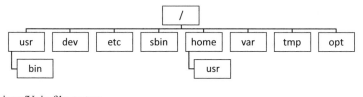

Fig. 6.1 Linux/Unix file system

6.3   The Command Line Tool

NGS analyses are performed in most cases using the terminal (Fig. 6.2). The command line interpreter *Bash* will translate and execute commands entered after the prompt ($). A basic overview of essential Unix/Linux commands that will allow you to navigate a file system and move, copy, edit files is provided in the Command line Bootcamp (http://korflab. ucdavis.edu/bootcamp.html) by Keith Bradnam. A beginner level Linux/Unix expertise is sufficient to get started and to perform the majority of analyses that we cover in this book.

The basic Linux/Unix commands, their meaning and description are depicted in Table 6.1. Each command is composed of three parts:

- The command.
- Options/ flags.
- Arguments.

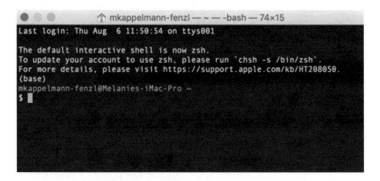

Fig. 6.2 The terminal using the CLI *Bash*

Table 6.1 Basic Linux/Unix commands

Help on any Linux/Unix command	
man [command]	Type **man rm** to read the manual for the **rm** command
[command] -h	Give short description of command
List a directory	
ls [path]	It is ok to combine attributes, eg **ls -laF** gets a long listing of all files with types
ls [path_1] [path_2]	List both [path_1] and [path_2]
ls -l [path]	Long listing, with date, size and permissions
ls -a [path]	Show all files, including important .dot files that do not otherwise show
ls -F [path]	Show type of each file. "/" = directory, "*" = executable
ls -R [path]	Recursive listing, with all subdirs
ls [path] l more	Show listing one screen at a time
Change to directory	
cd [dirname]	There must be a space between
cd ~	Go back to home directory, useful if you are lost
cd ..	Go back one directory
Make a new directory	
mkdir [dirname]	
Remove a directory	
rmdir [dirname]	Only works if [dirname] is empty
rm -r [dirname]	Remove all files and subdirs. **Careful!**
Print working directory	
pwd	Show where you are as full path. Useful if you are lost or exploring
Copy a file or directory	
cp [file1] [file2]	
cp -r [dir1] [dir2]	Recursive, copy directory and all subdirs
cat[newfile]>>[oldfile]	Append newfile to end of oldfile
Move (or rename) a file	
mv [oldfile] [newfile]	Moving a file and renaming it are the same thing
mv[oldname][newname]	
Delete a file	
rm [filespec]	**?** and * wildcards work like DOS should. "?" is any character; "*" is any string of characters
ls[filespec]rm [filespec]	Good strategy: first list a group to make sure it is what is you think......then delete it all at once
View a text file	
more [filename]	View file one screen at a time
less [filename]	Like **more**, with extra features
head [filename]	Command writes the first ten lines of a file to the screen

(continued)

Table 6.1 (continued)

Help on any Linux/Unix command	
tail [filename]	Command writes the last ten lines of a file to the screen
cat [filename]	View file, but it scrolls
cat [filename] | more	View file one screen at a time
gzcat or zcat [filename]	View a gziped file
open[filename/URL name/foldername]	To open a file, folder or URL
Edit a text file	
gedit [filename]	Basic text editor
sort [filename]	Sort line of text file
nano [filename]	Editing small text files
"Control + X"	Exits nano program
Create a text file	
cat > [filename]	Enter your text (multiple lines with **enter** are ok) and press **control-d** to save
gedit [filename]	Create some text and save it
touch [filename]	Creating empty files
Compare two files	
diff [file1] [file2]	Show the differences
sdiff [file1] [file2]	Show files side by side
Other text commands	
grep '[pattern]' [file]	Find regular expression in file
spell [file]	Display misspelled words
wc [file]	Count words in file
wc -l [file]	Count the number of lines in a file
Make an Alias	
alias [name]='[command]'	Put the command in 'single quotes'. More useful in your **.bashrc** file
Wildcards and Shortcuts	
*	Match any string of characters, e.g. **page*** gets page1, page10, and page.txt
?	Match any single character, e.g. **page?** gets page1 and page2, but not page10
[...]	Match any characters in a range, e.g. **page[1-3]** gets page1, page2, and page3
~	Short for your home directory, e.g. **cd ~** will take you home, and **rm -r ~** will destroy it
.	The current directory
..	One directory up the tree, eg **ls ..**
Pipes and Redirection	(You **pipe** a command to another command, and **redirect** it to a file.)
[command] > [file]	Redirect output to a file, e.g. **ls > list.txt** writes directory to file

(continued)

Table 6.1 (continued)

Help on any Linux/Unix command	
[command] >> [file]	Append output to an existing file, e.g. **cat update** >> **archive** adds update to end of archive
[command] < [file]	Get input from a file, eg **sort** < **file.txt**
[command] < [file1] > [file2]	Get input from file1, and write to file2, eg **sort** < **old.txt** > **new.txt** sorts old.txt and saves as new.txt
[command] | [command]	Pipe one command to another, e.g. **ls | more** gets directory and sends it to **more** to show it one page at a time
System info	
date	Show date and time
df	Check system disk capacity
du	Check your disk usage and show bytes in each directory
du -h	Check your disk usage in a human readable format
printenv	Show all environmental variables
uptime	Find out system load
w	Who is online and what are they doing?
top	Real time processor and memory usage
Others	
clear	Clear the terminal window
gzip or gunzip	GNU zip or unzip a file
sudo	Execute commands with administrative privileges
"Press up"	To retrieve previous commands
"Press down"	To go back
"/"	Is "root"
"~"	Is "HomeDir"
"|"	Pipe command to chain programs together
"Press TAB key"	Autocompletion of commands, files and directory names
q	Exit
top	Displays active processes. Press q to quit
Changing access rights	
chmod	Changing a file mode
	u	User
	g	Group
	o	Other
	a	all
	r	read
	w	write (and delete)
	x	execute (and access directory)
	+	add permission
	—	take away permission

Note that multiple flags can be set at once, e.g. `ls -alh` instead of the long version `ls -a -l -h`.

Executing commands from the terminal also needs certain filenames. Spaces within filenames are not accepted, as well as special characters and you always should use the specific file extension:

Good filenames:	project.txt	my_big_program.c	fred_dave.doc
Bad filenames:	project	my big program.c	fred & dave.doc

Take Home Message
- The command line is a powerful tool to navigate your file system, to explore and manipulate certain files.
- Commands have various additional options, which are documented for each kind of command (*help* or *man*).
- Different commands can be linked by building pipes via "|".
- Many NGS data analysis and text processing tasks can be achieved via a combination of simple UNIX programs piped together ("command line scripts").

Further Reading Mark G. Sobell. 2009. Practical Guide to Linux Commands, Editors, and Shell Programming, (2nd. ed.). Prentice Hall PTR, USA.

William E. Shotts. 2019. *The Linux command line: a complete introduction*, (2nd ed.). No Starch Press, USA.

The Biostar Handbook: 2nd Edition.

https://www.biostarhandbook.com/.

Review Questions
1. What are the main options for getting help about a particular command?
2. What are the options/ arguments in the call "ls -l –t"? What does it do?
3. Which options of the ls command can be used to list files sorted by their size? What additional option can be used to control that it works?
4. How can you search for a certain command in your *shell* History?
5. How can you combine different commands?

Answers to Review Questions

Answer to Question 1: --h or -h parameter of most programs; use the manual command man; use the info command -a more extensive documentation or Google it.

Answer to Question 2: List long listing format, sort by modification time.pt

Answer to Question 3: ls -l -S -s. 4. Ctrl R. 5. "Piping" |.

Acknowledgements We are grateful to Dr. Philipp Torkler (Senior Bioinformatics Scientist, *Exosome Diagnostics a Bio-Techne brand*, Munich, Germany) for critically reading this text. We thank for correcting our mistakes and suggesting relevant improvements to the original manuscript.

NGS Data

7

Melanie Kappelmann-Fenzl

Contents

M. Kappelmann-Fenzl (✉)
Deggendorf Institute of Technology, Deggendorf, Germany

Institute of Biochemistry (Emil-Fischer Center), Friedrich-Alexander University Erlangen-Nürnberg, Erlangen, Germany
e-mail: melanie.kappelmann-fenzl@th-deg.de

© Springer Nature Switzerland AG 2021
M. Kappelmann-Fenzl (ed.), *Next Generation Sequencing and Data Analysis*, Learning Materials in Biosciences, https://doi.org/10.1007/978-3-030-62490-3_7

What You Will Learn in This Chapter
Next-generation sequencing experiments produce millions of short reads per sample and the processing of those raw reads and their conversion into other file formats lead to additional information on the obtained data. Various file formats are in use in order to store and manipulate this information. This chapter presents an overview of the file formats FASTQ, FASTA, SAM/BAM, GFF/GTF, BED, and VCF that are commonly used in analysis of next-generation sequencing data. Moreover, the structure and function of the different file formats are reviewed. This chapter explains how different file formats can be interpreted and what information can be gained from their analysis.

7.1   Introduction

Analyzing NGS data means to handle really big data. Thus, one of the most important things is to store all these "big data" in appropriate data formats to make them manageable. The different data formats store in many cases the same type "of information, but not all data formats are suitable for all bioinformatic tools. Always keep in mind that bioinformatic software tools do not transform data into answers, they transform data into other data formats. The answers result from investigating, understanding, and interpreting the various data formats.

7.2   File Formats

7.2.1   Basic Notations

- *Fragment:* The molecule to be sequenced.
- *Read:* One sequenced part of a biological fragment (mate I or mate II).
- *Mate I:* The sequence of the 5'end of a paired-end sequencing approach.
- *Mate II:* The sequence of the 3'end of a paired-end sequencing approach.
- *Sequencing depth:* The number of all the sequences, reads, or bases represented in a single sequencing experiment divided by the target region.
- *Sequencing Coverage:*

    The theoretical redundancy of coverage (c) is described as LN/G, where L is the read length, N is the number of reads, and G is the haploid genome length [1].

    Sequencing coverage can be calculated in different ways depending on reference points (whole genome, one locus, or one position in the genome):

1. One locus: # of bases mapping to the locus/size of locus.

Table 7.1 Sequencing coverage recommendations for some common sequencing methods

Sequencing method	Recommended coverage
Whole genome sequencing	$\sim 30\times$ to $50\times$ (for human)
Whole-exome sequencing	$\sim 100\times$
RNA sequencing	~ 20–50 Mio. reads/sample
ChIP-Seq	$\sim 100\times$

2. One position: # of reads overlapping with one position.
3. Whole genome: # of sequenced bases/size of genome.

The necessary sequencing coverage strongly depends on the performed sequencing method (WGS, WES, RNA-Seq, ChIP-Seq), the reference genome size, and gene expression patterns. Recommendations of sequencing coverage regarding the sequencing method are listed in Table 7.1 ("Sequencing Coverage". illumina.com. Illumina education. Retrieved 2017-10-02.).

Review Question 1

What coverage of a human genome will one get with one lane of HiSeq3000 in paired-end Sequencing mode (2x75), if 300M clusters were bound (human genome size: 3.2 GB)?

7.2.2 FASTA

FASTA format is a text-based format for representing either nucleotide sequences or peptide sequences, in which nucleotides or amino acids are represented using a single-letter code. The simplicity of FASTA format makes it easy to manipulate and parse using text-processing tools and scripting languages like R, Python, Ruby, and Perl. A sequence in FASTA format begins with a single-line description, followed by lines of sequence data. The so-called defline starts with a ">" symbol and can thus be distinguished from the sequence data. A more detailed description of the FASTA format and its purpose within BLAST search can be found on the NCBI website https://blast.ncbi.nlm.nih.gov/Blast.cgi?CMD=Web&PAGE_TYPE=BlastDocs&DOC_TYPE=BlastHelp.

FASTA format example:

```
> Sequence XY description
CACAGCCAGCCAGCCAGGTCGGCAGTATAGTCCGAACTGCAAATCTTATTTTCTTTTCACCTTCTCTCTAA
CTGCCCAGAGCTAGCGCCTGTGGCTCCCGGGCTGGTGTTTCGGGAGTGTCCAGAGAGCCTGGTCTCCAGCC
GCCCCCGGGAGGAGAGCCCTGCTGCCCAGGCGCTGTTGACAGCGGCGGAAAGCAGCGGTACCCACGCGCCC
GCCGGGGGGAAGTCGGCGAGCGGCTGCAGCAGCAAAGAACTTTCCCGGCTGGGAGGACCGGAGACAA
```

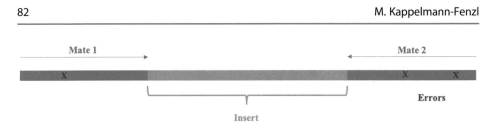

Fig. 7.1 Sequencing read (ends of DNA fragment for mate pairs)

7.2.3   FASTQ

Once you have sequenced your samples of interest you get back your sequencing data in a certain format storing the reads information. Each sequencing read (i.e., paired-end) is structured as depicted in Fig. 7.1. All reads and their information are stored in a file format called FASTQ.

Sequencing facilities often store the read information in *.*fastq* or unaligned *.*bam* files. The latter can be transformed in a *.*fastq* file via *BEDTools*:

```
# in terms of paired end sequencing data:
bedtools bamtofastq -I *.bam -fq *_r1.fastq -fq2 *_r2.fastq
# in terms of single end sequencing data:
bedtools bamtofastq -I *.bam -fq *.fastq
```

The FASTQ format is a text-based standard format for storing both, a DNA sequence and its corresponding quality scores from NGS. There are four lines per sequencing read.

FASTQ format example:

```
@HWI-K00288_BSF_0436:4:1101:10003:10669 length=76
CTCCATCAGGTTTGCCCACTGGCGATGCATGTCTTCCACTATTGGCTCGAACCAGCCCTTGAGGCGGGCCTGGAAG
+
AAF<<AJJJJJJFJJJJJF<JFFFFJJJJJJFJJJJJFJJFJJJFJJFJJPAJJJFAAJJFAAJ<JA
APJJJJFJJJFJAF
```

- The first line starts with '@', followed by the label.

- The second line represents the sequence of the read.

- The third line starts with '+', serving as a separator.

- The fourth line contains the Q scores (quality values for sequence in line 2) represented as ASCII characters

Dec	Hex	Char	Dec	Hex	Char	Dec	Hex	Char	Dec	Hex	Char
0	00	Null	32	20	Space	64	40	@	96	60	`
1	01	Start of heading	33	21	!	65	41	A	97	61	a
2	02	Start of text	34	22	"	66	42	B	98	62	b
3	03	End of text	35	23	#	67	43	C	99	63	c
4	04	End of transmit	36	24	$	68	44	D	100	64	d
5	05	Enquiry	37	25	%	69	45	E	101	65	e
6	06	Acknowledge	38	26	&	70	46	F	102	66	f
7	07	Audible bell	39	27	'	71	47	G	103	67	g
8	08	Backspace	40	28	(	72	48	H	104	68	h
9	09	Horizontal tab	41	29	)	73	49	I	105	69	i
10	0A	Line feed	42	2A	*	74	4A	J	106	6A	j
11	0B	Vertical tab	43	2B	+	75	4B	K	107	6B	k
12	0C	Form feed	44	2C	,	76	4C	L	108	6C	l
13	0D	Carriage return	45	2D	-	77	4D	M	109	6D	m
14	0E	Shift out	46	2E	.	78	4E	N	110	6E	n
15	0F	Shift in	47	2F	/	79	4F	O	111	6F	o
16	10	Data link escape	48	30	0	80	50	P	112	70	p
17	11	Device control 1	49	31	1	81	51	Q	113	71	q
18	12	Device control 2	50	32	2	82	52	R	114	72	r
19	13	Device control 3	51	33	3	83	53	S	115	73	s
20	14	Device control 4	52	34	4	84	54	T	116	74	t
21	15	Neg. acknowledge	53	35	5	85	55	U	117	75	u
22	16	Synchronous idle	54	36	6	86	56	V	118	76	v
23	17	End trans. block	55	37	7	87	57	W	119	77	w
24	18	Cancel	56	38	8	88	58	X	120	78	x
25	19	End of medium	57	39	9	89	59	Y	121	79	y
26	1A	Substitution	58	3A	:	90	5A	Z	122	7A	z
27	1B	Escape	59	3B	;	91	5B	[	123	7B	{
28	1C	File separator	60	3C	<	92	5C	\	124	7C	|
29	1D	Group separator	61	3D	=	93	5D	]	125	7D	}
30	1E	Record separator	62	3E	>	94	5E	^	126	7E	~
31	1F	Unit separator	63	3F	?	95	5F	_	127	7F	⌂

Example:

!+EI

ASCII	33	43	69	73
-33*				
Phred	0	10	36	40

Phred Quality Score	Probability of incorrect base call	Base Call Accuracy
10	1 in 10	90%
20	1 in 100	99%
30	1 in 1.000	99.9%
40	1 in 10.000	99.99%
50	1 in 100.000	99.999%

*To make Phred Score zero based.
Source: www.shsu.edu

Fig. 7.2 ASCII table and Phred Score calculation

HWI-K00288_BSF_0436	Instrument name/ID
4	Flowcell lane
1101	Tile number within the flowcell lane
10003	"x"-coordinate of the cluster within the tile
10669	"y"-coordinate of the cluster within the tile

The Phred Score represents the probability that the corresponding base call is incorrect [2].

The ASCII table and an example for Phred Score calculation are depicted in Fig. 7.2. Qualities are based on the Phred scale and are *encoded*: $Q = -10*\log_{10}(P_{err})$.

The formula for getting Phred quality from *encoded* quality is: $Q = ascii(char) - 33$.

Review Question 2

Determine the Phred-Scores for the following quality values in a FASTQ file format: #4=DBDDDHFHFFHIGHIII.

7.2.4 SAM

SAM stands for Sequence Alignment/Map format. As the name suggests, you get this file format after mapping the fastq files to a reference genome. It is a TAB-delimited text format

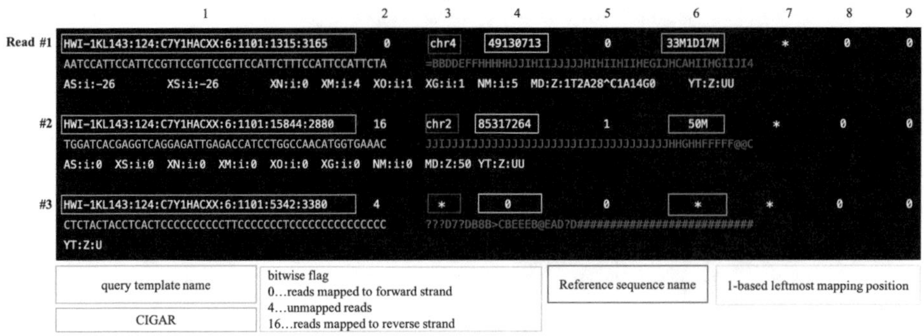

Fig. 7.3 SAM format specification of a 50 bp single-end sequencing approach

consisting of a header section, which is optional but strongly encouraged to include, and an alignment section. The header lines start with "@," while alignment lines do not.

Each alignment line has 11 mandatory fields for essential alignment information such as mapping position, and also has variable number of optional fields for flexible or aligner specific information [3] (Fig. 7.3, Table 7.2).

In the alignment section of a SAM file, CIGAR (Concise Idiosyncratic Gapped Alignment Report) is one of the 11 mandatory fields of each alignment line. The CIGAR string is a sequence of base lengths and associated operations. CIGAR string is used to indicate whether there is a match or mismatch between the bases of read and the reference sequence. It is quite useful for detecting insertions or deletions (Indels). The CIGAR string is read based and there are two different versions. In the following example POS5 indicates the starting position of read alignment to the reference.

Example:

RefPos:	1	2	3	4	5	6	7		8	9	10	11	12	13	14	15	16	17
Reference:	C	C	A	T	A	C	T		G	A	A	C	T	G	A	C	T	A
Read:					A	C	T	A	G	A	A		T	G	G	C	T	

POS: 5: Indicating the starting position of read alignment on the reference.

Version 1

The CIGAR string of this version does not distinguish between a match and a mismatch (see Position 14 in the example above). Consequently, to calculate the number of errors for an alignment, the CIGAR string and the MD-TAG (Optional field) are mandatory.

Note: A MD-TAG String encodes mismatched and deleted bases and is reference-based. The MD-TAG string *MD : Z: 7^C5* means from the leftmost reference base in the

Table 7.2 Description of TAGs in the SAM file format depicted in Fig. 7.3

	@HD	The header line; VN: Format version; SO: Sorting order of alignments.
	@SQ	Reference sequence dictionary.
	@RG	Read group information.
	@PG	Program ID: Program record identifier; VN: Program version; CL: Command line
1	QNAME	Query template name. Used to group/identify alignments that are together, like paired alignments or a read that appears in multiple alignments.
2	FLAG	Bitwise Flag. Bitwise set of information describing the alignment by answering multiple questions. Decoding of the bitwise flag can be performed here: http://broadinstitute.github.io/picard/explain-flags.html
3	RNAME	Reference sequence name (e.g. Chromosome name).
4	POS	Leftmost position of where this alignment maps to the reference. For SAM, the reference starts at 1, so this value is 1-based. [For BAM the reference starts at 0, so this value is 0-based.]
5	MAPQ	Mapping quality.
6	CIGAR	String indicating alignment information that allows the storing of clipped. Old Version:
7	RNEXT	The reference sequence name of the next alignment in this group.
8	PNEXT	Leftmost position of where the next alignment in this group maps to the reference.
9	TLEN	Length of this group from the leftmost position to the rightmost position.
10	SEQ	The query sequence for this alignment.
11	QUAL	Query quality for this alignment (one for each base in the query sequence).
	Optional field	Additional optional information is also contained within the alignment in in TAG:TYPE:VALUE format.

alignment, there are 7 matches followed by a deletion from the reference, whereas the deleted base is *C* and the last 5 bases are matches.

  Example CIGAR: 3M1I3M1D5M

3M   3 matches/mismatches
1I   1 insertion
3M   3 matches/mismatches
1D   1 deletion
5M   5 matches/mismatches

  Version 2 (new Version)

Table 7.3 CIGAR string for error calculation (optional field: NM:i:<#>)

<#> Operator	CIGAR string description (<#> means *number of*)
<#>=	<#> of matches
<#>X	<#> of mismatches
<#>D	<#>Deletions (gap in the sequencing read)
<#>I	<#>Insertions (gap in the reference sequence)
<#>N	<#> skipped region (gap in the sequencing read)
S	Soft clipping (clipped sequences present in SEQ)
H	Hard clipping (clipped sequences not present in SEQ)

Here the CIGAR string does distinguish between matches and mismatches, thus the MD-TAG is not mandatory. In general, a CIGAR string is made up of <integer> and <op>, where <op> is an operation specified as a single character (Table 7.3).

Example CIGAR: 3=1I3=1D2=1X2=

7.2.5 BAM

A BAM file (*.*bam*) is the compressed binary version of a SAM file. BAM files are binary files, which mean they cannot be opened like text files; they are compressed and can be sorted and/or indexed.

They consist of a header section and an alignment section. The header contains information about the entire file, such as sample name and length. Alignments contain the name, sequence, and quality of a read. Alignment information and custom tags can also be found in the alignment section.

The following table shows the information for each read or read pair depicted in the alignment section:

RG	Read group, which indicates the number of reads for a specific sample
BC	Barcode tag, indicating the read-associated demultiplexed sample ID
SM	Single-end alignment quality
AS	Paired-end alignment quality
NM	Edit distance tag, recording the Levenshtein distance between read and reference
XN	Amplicon name tag, recording the amplicon tile ID associated with the read

BAM files can also have a companion file, called an index file. This file has the same name, suffixed with *.*bai*. The BAI file acts like an external table of contents, and allows programs to jump directly to specific parts of the BAM file without reading through all of the sequences. Without the corresponding BAM file, your BAI file is useless, since it does not actually contain any sequence data.

Thus, if you want to visualize your BAM file, in the IGV Browser [4–6], for example, you need the corresponding BAI file to do so. Another possibility to visualize your BAM alignment file is to upload your files to the *bam.iobio* online tool (https://bam.iobio.io/).

To convert a SAM file format into a BAM file format you can use samtools and then sort your obtained BAM file by coordinates. Often these steps are already included in your mapping job, if not, you can easily run the following commands on your SAM file to do so:

```
#convert SAM to BAM
samtools view -S -b *.sam > *.bam
# sort BAM file via samtools sort
samtools sort *bam -o *.sorted.bam
#index the sorted BAM file
Samtools index *.sorted.bam
```

7.2.6 GFF/GTF

GFF stands for *General Feature Format* and GTF for *Gene Transfer Format*. Both are annotation files. An annotation can be thought of as a label applied to a region of a molecule. The GFF/GTF formats are 9 column tab-delimited formats. Every single line represents a region on the annotated sequence and these regions are called features. Features can be functional elements (e.g., genes), genetic polymorphisms (e.g., SNPs, INDELs, or structural variants), or any other annotations. Each feature should have a type associated. Examples of some possible types are: SNPs, introns, ORFs, UTRs, etc. In the GFF format both the start and the end of the features are 1-based.

The GTF format is identical to the second version of GFF format. In terms of the 3rd version GFF format the first eight GTF fields are the same, but the layout of the 9th (last) column of the GTF is different. The feature field is the same as GFF, with the exception that it also includes the following optional values: 5'UTR, 3'UTR, inter, inter_CNS, and intron_CNS. The group field has been expanded into a list of attributes. Each attribute consists of a type/value pair. Attributes must end in a semi-colon, and be separated from any following attribute by exactly one space.

The attribute list must begin with the two mandatory attributes:

- gene_id value—A globally unique identifier for the genomic source of the sequence.
- transcript_id value—A globally unique identifier for the predicted transcript.

TAB-separated standard GTF columns are:

Column #	Content	Values/Format
1	Chromosome name	chr [1,2,3,4,5,6,7,8,9,10,11,12,13,14,15,16,17,18,19,20,21,22, X,Y,M] or GRC accession [a]
2	Annotation source	[ENSEMBL,HAVANA]
3	Feature type	[gene,transcript,exon,CDS,UTR,start_codon,stop_codon, Selenocysteine]
4	Genomic start location	Integer-value (1-based)
5	Genomic end location	Integer-value
6	Score (not used)	.
7	Genomic strand	[+,−]
8	Genomic phase (for CDS features)	[0,1,2,.]
9	Additional information as key-value pairs	see https://www.gencodegenes.org/pages/data_format. html

GTF/GFF files, as well as FASTA files can be downloaded from databases like GENCODE (https://www.gencodegenes.org/), ENSEMBL (https://www.ensembl.org/downloads.html), UCSC (http://hgdownload.cse.ucsc.edu/downloads.html), etc. You need those file formats for generation of a genome index together with the corresponding FASTA file of the genome as it is described earlier in this Chapter in Sect. 7.2.2.

A more detailed description about GFF/GTF file formats can be found on https://www.ensembl.org/info/website/upload/gff.html and many other websites, which make these available for download.

Review Question 3

Why do we have to download *.fa and *.gtf/*.gff of a certain genome of interest?

7.2.7   BED

The BED format provides a simpler way of representing the features in a molecule. Each line represents a feature in a molecule and it has only three required fields: name (of chromosome or scaffold), start, and end. The BED format uses 0-based coordinates for the starts and 1-based for the ends. Headers are allowed. Those lines should be preceded by # and they will be ignored.

The first three columns in a BED file are required, additional columns are optional [7, 8].

If you display the first lines of a BED file in the terminal, it looks like this:

```
#chrom    chromStart    chromEnd
chr1      1213941196    213942363
chr1      213942363     213943530
chr1      213943530     213944697
chr2      158364697     158365864
chr2      158365864     158367031
chr3      127477031     127478198
chr3      127478198     127479365
chr3      127479365     127480532
chr3      127480532     127481699
```

For further information about BED files see /https://www.ensembl.org/info/website/upload/bed.html/.

The GFF/GTF/BED formats are the so-called interval formats that retain only the coordinate positions for a region in a genome. A genome interval sequencing data format can describe more or less all genetic structures, alterations, variants, etc.:

- Genes: exons, introns, UTRs, promoters
- Conservation
- Genetic variation
- Transposons
- Origins of replication
- TF binding sites
- CpG islands
- Segmental duplications
- Sequence alignments
- Chromatin annotations
- Gene expression data
- And many more

Due to the fact, that we are handling intervals, many complex analyses can be reduced to genome arithmetic. Sounds complicated, but actually all you need are some basic mathematical operations like addition, subtraction, multiplication, and division. Therefore, some very clever bioinformaticians developed a tool for genome "calculations"—*BEDTools* (https://bedtools.readthedocs.io/en/latest/). With this tool you can easily answer the following questions by comparing two or more BED/BAM/VCF/GFF files (Fig. 7.4):

- Which gene is the closest to a ChIP-seq peak?
- Is my latest discovery novel?
- Is there strand bias in my data?

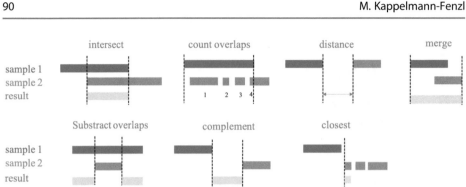

Fig. 7.4 *BEDTools* commands and their results by comparing to different samples

- How many genes does this mutation affect?
- Where did I fail to collect sequence coverage?
- Is my favorite feature significantly correlated with some other feature?

To ensure a safe handling with the program the following tutorial is recommended: http://quinlanlab.org/tutorials/bedtools/bedtools.html.

7.2.8   BedGraph

The BedGraph (**.bg*) format is based on the BED format with a few differences and allows display of continuous-valued data in track format. This display type is useful for probability scores and transcriptome data. The data are preceded by a track definition line, which adds a number of options for controlling the default display of this track. The fourth column of this file format provides information about regions of the genome with sufficient read coverage [9]. Thus, after converting this format into a bigWig (a binary indexed version) it is suitable for visualizing sequencing data in the UCSC Genome Browser (https://genome.ucsc.edu/).

track	type=bedGraph	name="BedGraph Format"	description="BedGraph	format"	priority=20
chr19	59302000	59302300	-1.0		
chr19	59302300	59302600	-0.75		
chr19	59302600	59302900	-0.50		
chr19	59302900	59303200	-0.25		
chr19	59303200	59303500	0.0		
chr19	59303500	59303800	0.25		
chr19	59303800	59304100	0.50		
chr19	59304100	59304400	0.75		

```
##fileformat=VCFv4.2
##fileDate=20090805
##source=myImputationProgramV3.1
##reference=file:///seq/references/1000GenomesPilot-NCBI36.fasta
##contig=<ID=20,length=62435964,assembly=B36,md5=f126cdf8a6e0c7f379d618ff66beb2da,species="Homo sapiens",taxonomy=x>
##phasing=partial
##INFO=<ID=NS,Number=1,Type=Integer,Description="Number of Samples With Data">
##INFO=<ID=DP,Number=1,Type=Integer,Description="Total Depth">                      ## meta-information
##INFO=<ID=AF,Number=A,Type=Float,Description="Allele Frequency">
##INFO=<ID=AA,Number=1,Type=String,Description="Ancestral Allele">                  # header line
##INFO=<ID=DB,Number=0,Type=Flag,Description="dbSNP membership, build 129">
##INFO=<ID=H2,Number=0,Type=Flag,Description="HapMap2 membership">                  data lines
##FILTER=<ID=q10,Description="Quality below 10">
##FILTER=<ID=s50,Description="Less than 50% of samples have data">
##FORMAT=<ID=GT,Number=1,Type=String,Description="Genotype">
##FORMAT=<ID=GQ,Number=1,Type=Integer,Description="Genotype Quality">
##FORMAT=<ID=DP,Number=1,Type=Integer,Description="Read Depth">
##FORMAT=<ID=HQ,Number=2,Type=Integer,Description="Haplotype Quality">
#CHROM POS     ID        REF   ALT    QUAL FILTER INFO                              FORMAT      NA00001            NA00002            NA00003
20     14370   rs6054257 G     A      29   PASS   NS=3;DP=14;AF=0.5;DB;H2           GT:GQ:DP:HQ 0|0:48:1:51,51 1|0:48:8:51,51 1/1:43:5:.,.
20     17330   .         T     A      3    q10    NS=3;DP=11;AF=0.017              GT:GQ:DP:HQ 0|0:49:3:58,50 0|1:3:5:65,3   0/0:41:3
20     1110696 rs6040355 A     G,T    67   PASS   NS=2;DP=10;AF=0.333,0.667;AA=T;DB GT:GQ:DP:HQ 1|2:21:6:23,27 2|1:2:0:18,2   2/2:35:4
20     1230237 .         T     .      47   PASS   NS=3;DP=13;AA=T                   GT:GQ:DP:HQ 0|0:54:7:56,60 0|0:48:4:51,51 0/0:61:2
20     1234567 microsat1 GTC   G,GTCT 50   PASS   NS=3;DP=9;AA=G                    GT:GQ:DP    0/1:35:4       0/2:17:2       1/1:40:3
```

Fig. 7.5 The general structure of VCF format (modified according to https://samtools.github.io/hts-specs/VCFv4.2.pdf)

7.2.9 VCF

The VCF (*Variant Call Format*) contains information about genetic variants found at specific positions in a reference genome. The VCF header includes the VCF file format version and the variant caller version. The header lists the annotations used in the remainder of the file. The VCF header includes the reference genome file and BAM file. The last line in the header contains the column headings for the data lines (Fig. 7.5) [10].

VCF File Data Lines—Each data line contains information about a single variant. VCF Tools is a program designed for working with VCF files and can be used to perform the following operations on VCF files:

- Filter out specific variants.
- Compare files.
- Summarize variants.
- Convert to different file types.
- Validate and merge files.
- Create intersections and subsets of variants.

For example:

Each data line contains an information about a certain position in the genome. The example above shows:

1. A SNP (G→A) with a quality of 29.
2. A possible SNP (T→A) that has been filtered out because its quality is below 10.
3. A site at which two alternate alleles are called.
4. A site that is called monomorphic reference (i.e., with no alternate alleles).
5. A microsatellite with two alternative alleles, one a deletion of 2 bases (TC), and the other an insertion of one base (T).

You can have a look at a generated VCF and source file (i.e. *.*bam*) in the IGV Browser [4–6]. You can find further information on VCF files in Chap. 10 and at https://samtools. github.io/hts-specs/VCFv4.2.pdf.

7.2.10 SRA (Sequence Read Archive)

The Sequence Read Archive (SRA) is a bioinformatics database from NCBI (National Center for Biotechnology Information) that provides a public repository for sequencing data, generated by high-throughput sequencing. The SRA-toolkit (https://github.com/ncbi/ sra-tools) is needed to download the data [11].

The following command can be used to download an SRA file of a paired-end sequencing experiment and to store mate I in *_*I.fastq* and mate II in *_*II.fastq*. The – gzip option is used to minimize the size of the two fastq files.

```
fastq-dump –split-3 –gzip SRR[number]
```

In terms of a single-end sequencing experiment you would type:

```
fastq-dump –gzip SRR[number]
```

Review Question 4

Find the sample with the accession SRR10257831 from the Sequence Read Archive and find out the following information:

- Which species?
- Genome or transcriptome?
- What sequencing platform was used?
- What read length?
- Was it a single-end or paired-end sequencing approach?

7.3   Quality Check and Preprocessing of NGS Data

7.3.1   Quality Check via *FastQC*

FASTQ files (see Sect. 7.2.3) are the "raw data files" of any sequencing application, that means they are "untouched." Thus, this file format is used for Quality Check of sequencing reads. The Quality Check procedure is commonly done with the *FastQC* tool written by Simon Andrews of Babraham Bioinformatics (https://www.bioinformatics.babraham.ac. uk/projects/fastqc/). *FastQC* and other similar tools are useful for assessing the overall quality of a sequencing run and are widely used in NGS data production environments as an initial QC checkpoint [12]. This tool provides a modular set of analyses which you can use to give a quick impression of whether your data has any problems of which you should be aware before doing any further analysis.

The main features of *FastQC* are:

- Import of data from *.bam*, *.sam*, or *.fastq* files (any variant).
- Providing a quick overview to tell you in which areas there may be problems. In a perfect world your *FastQC* report would look like this:

        ✅ Basic Statistics
        ✅ Per base sequence quality
        ✅ Per tile sequence quality
        ✅ Per sequence quality scores
        ✅ Per base sequence content
        ✅ Per sequence GC content
        ✅ Per base N content
        ✅ Sequence Length Distribution
        ✅ Sequence Duplication Levels
        ✅ Overrepresented sequences
        ✅ Adapter Content

But unfortunately, we do not live in a perfect world and therefore it will rarely happen that you receive such a report. *FastQC* reports include summary graphs and tables to quickly assess your data.

You can run the *FastQC* program from the terminal

```
fastqc *.fastq_1 *.fastq_2 [you can add as many files as you want] -o path/to/
outputdir
#if you do not want to type in every single fastq file name you can also run
find . -name "*.fastq" | parallel fastqc -o {//}/{}
#in the current working directory where the fastq files are located
# (if your fastq files are in a fastq.gz format, what is highly recommended then
use the .fastq.gz #file extension)
```

On the top of the obtained *FastQC* HTML report a summary of the modules which were run, and a quick evaluation of whether the results of the module seem entirely normal (green tick), slightly abnormal (orange triangle), or very unusual (red cross) is shown.

In detail you will get graphs of all the modules mentioned above, which give you the information of your input data quality:

7.3.1.1 The Basic Statistics Module

This module represents simple information about input FASTQ file: its name, type of quality score encoding, total number of reads, read length, and GC content. Including:

- Filename: The original filename of the file which was analyzed.
- File type: Says whether the file appeared to contain actual base calls or colorspace data which had to be converted to base calls.
- Encoding: Says which ASCII encoding of quality values was found in this file.
- Total Sequences: A count of the total number of sequences processed. There are two values reported, actual and estimated.
- Filtered Sequences: If running in Casava mode sequences flagged to be filtered will be removed from all analyses. The number of such sequences removed will be reported here. The total sequences count above will not include these filtered sequences and will the number of sequences actually used for the rest of the analysis.
- Sequence Length: Provides the length of the shortest and longest sequence (long-read sequencing) in the set. If all sequences are the same length, only one value is reported (short-read sequencing).
- %GC: The overall %GC of all bases in all sequences.

7.3.1.2 Per Base Sequence Quality

This box-and-whisker plot shows the range of quality values (Phred-Scores; see Sect. 7.2.3) across all bases at each position in the FASTQ file.

The central red line is the median value, the yellow box represents the inter-quartile range (25–75%), the upper and lower whiskers represent the 10% and 90% points, and the blue line represents the mean quality.

On the x-axis the bases 1–10 are reported individually, then the bases are summarized in bins. The number of base positions binned together depends on the length of the read, thus shorter reads will have smaller windows and longer reads larger windows. The y-axis depicts the Phred-Scores.

Often you will see a decreasing quality with increasing base position (Figs. 7.6 and 7.7). This effect lies in the sequencing by synthesis technology of Illumina and is called *phasing*. During each sequencing cycle chemicals that include variants for all four nucleotides are washed over the flow cell. The nucleotides have a terminator cap so that only 1 base gets incorporated. After the detection of the fluorescence signal the terminator cap is removed and the next cycle can start. Accordingly, a synchronous sequencing of DNA fragments in each cluster by expressing specific fluorescence signals is guaranteed (see Sect. 4.2). The main reason for the decreasing sequence quality is that the blocker of a nucleotide is not correctly removed after signal detection (*phasing*) and thus lead to light pollution during signal detection. This error occurs more often over time and thus with an increasing read length.

7.3.1.3 Per Tile Sequence Quality

The Per tile Sequence Quality Graph graph only appears in your *FastQC* report if you are using an Illumina library. The original sequence identifiers are retained encoding the flowcell tile from which each read came (Fig. 7.8). Reasons for seeing errors on this plot could be transient problems such as bubbles going through the flow cell, or there could be

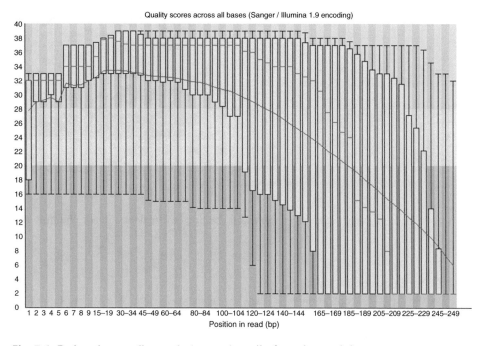

Fig. 7.6 Bad per base quality graph. (source: https://rtsf.natsci.msu.edu/)

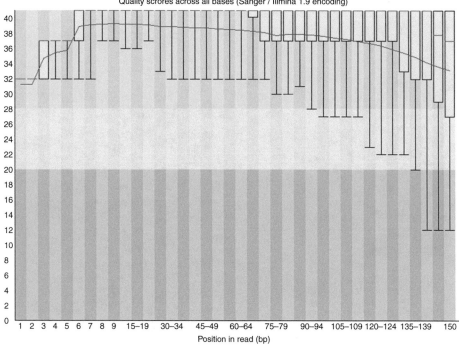

Fig. 7.7 Good per base quality graph. (source: https://rtsf.natsci.msu.edu/)

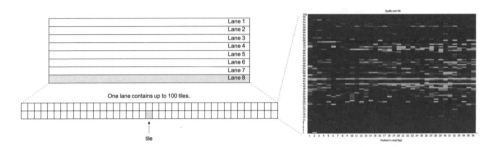

Fig. 7.8 Per Tile Sequence Quality. Schematic representation of a tile (left) and *FastQC* output (right). (source: https://www.bioinformatics.babraham.ac.uk)

more permanent problems such as smudges on the flowcell or debris inside the flow cell lane.

7.3.1.4 Per Sequence Quality Scores

This plot illustrates the total number of reads versus the mean sequence quality score over each full-length read and allows to see if a subset of your sequences has an universally poor quality (Fig. 7.9). For further data analysis steps only a small percentage of the total sequences should show a low quality (Fig. 7.10).

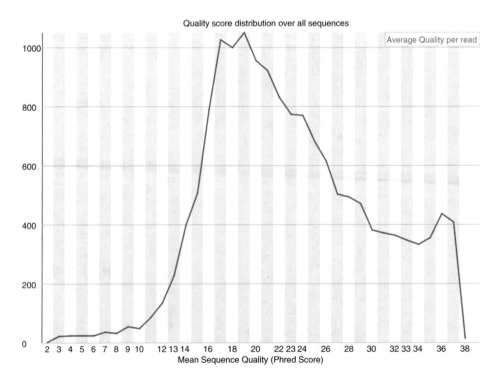

Fig. 7.9 Bad per sequence quality score. (source: https://rtsf.natsci.msu.edu/)

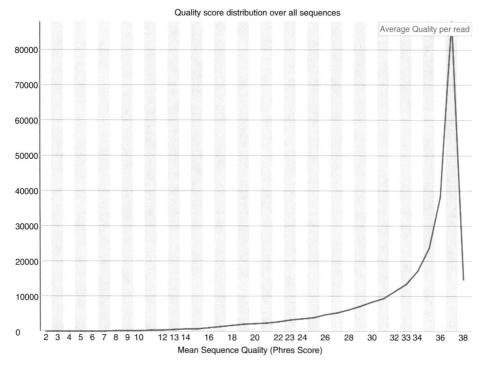

Fig. 7.10 Good per sequence quality score. (source: https://rtsf.natsci.msu.edu/)

7.3.1.5 Per Base Sequence Content

The plots below represent the percent of bases called for each of the four nucleotides (A/T/ G/C) at each position across all reads in the FASTQ file. Again, the x-axis is not uniform as described for Per base sequence quality (see Sect. 7.3.1.2). In a random sequencing library, you would expect that there would be almost no difference between the different bases of a sequence run, so the lines in this plot should run parallel to each other. If strong biases are detected which change in different bases this can usually be associated with a contamination of your library with overrepresented sequences, like clonal reads or adapters. Note that in DNA-Seq libraries the proportion of each base remains relatively constant over the length of a read (Fig. 7.11); however, most RNA-Seq libraries show a not uniform distribution of bases for the first 10–15 nucleotides. This is normal and expected (Fig. 7.12).

7.3.1.6 Per Base GC Content

The per base GC content plots the GC content of each base position in a file. In a random library the line in this plot should run horizontally. A consistent bias across all bases indicates that the original library was sequence biased or indicates a systematic problem during the sequencing run. A GC bias changing in different bases rather indicates a contamination with overrepresented sequences.

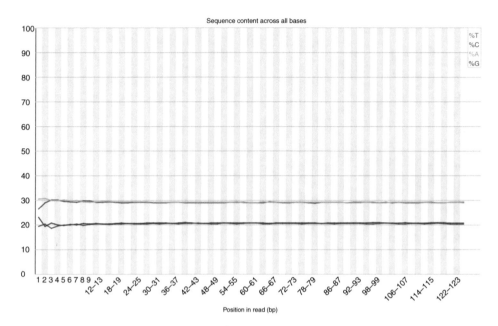

Fig. 7.11 Per base sequence content of DNA library. (source: https://rtsf.natsci.msu.edu/)

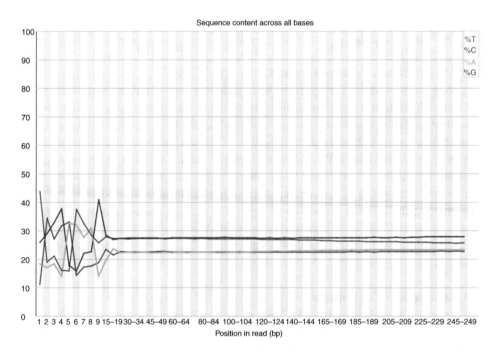

Fig. 7.12 Per base sequence content of RNA library. (source: https://rtsf.natsci.msu.edu/)

7.3.1.7 Per Sequence GC content

This plot illustrates the number of reads versus the GC content per read in percent. In terms of DNA sequencing all reads should form a normal distribution and the peak should be positioned at the mean GC content for the sequenced organism. In RNA sequencing approaches there may be a greater or lesser distribution of mean GC content among transcripts as it is depicted in Fig. 7.13. A shifted normal distribution indicates some systematic bias independent of base position.

7.3.1.8 Per Base N Content

This plot depicts the percentage of bases at each position or bin in a read with no base call ("N"). If the curve of the graph rises at any position noticeably above zero indicates a problem during the sequencing run. The report depicted in Fig. 7.14 the sequencing instrument was unable to call a base for round about 20% of the reads at position 29. In most cases a low proportion of Ns appear near the end of a sequence.

7.3.1.9 Sequence Length Distribution

In terms of short-read sequencing fragments of uniform length should be generated, depending on your sequencing settings (50 bp, 75 bp, 100 bp, 150 bp). However, this

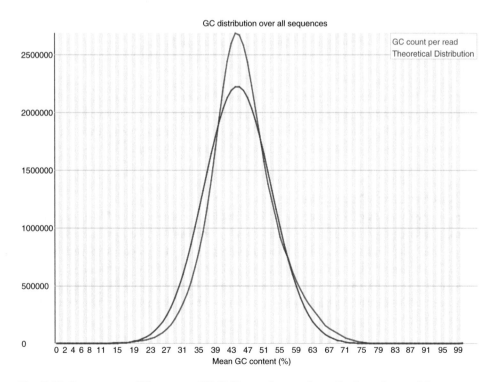

Fig. 7.13 Per sequence GC content of RNA library. (source: https://rtsf.natsci.msu.edu)

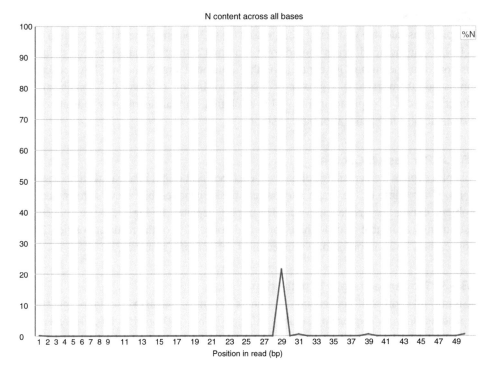

Fig. 7.14 Per base N content. (source: https://rtsf.natsci.msu.edu)

length can of course change after trimming sequences from the end due to poor quality base calls or adapter contamination (see Sect. 7.3.2). If you performed long-read sequencing, you will obtain a distribution of various read lengths, which means some reads are shorter, most reads have an enriched size distribution, and some are longer.

7.3.1.10 Sequence Duplication Levels

There are two sources of duplicate reads: PCR duplication due to biased PCR enrichment or really overrepresented sequences such as very abundant transcripts in an RNA-Seq library. PCR duplicates misrepresent the true proportion of sequences in your starting material, whereas really overrepresented sequences do faithfully represent your input. Thus, in DNA-Seq nearly 100% of your reads should be unique (Fig. 7.15), in RNA-Seq duplicate reads of highly abundant transcripts will be observed, however the duplication is normal in this case (Fig. 7.16).

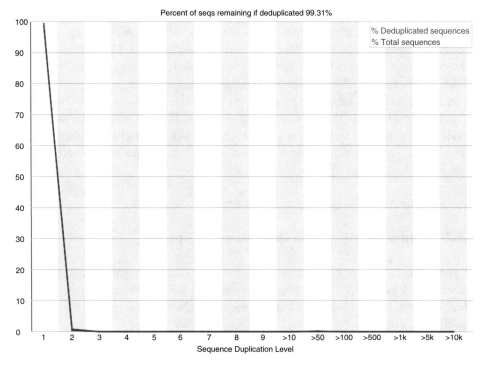

Fig. 7.15 DNA library sequence duplication. (source: https://rtsf.natsci.msu.edu)

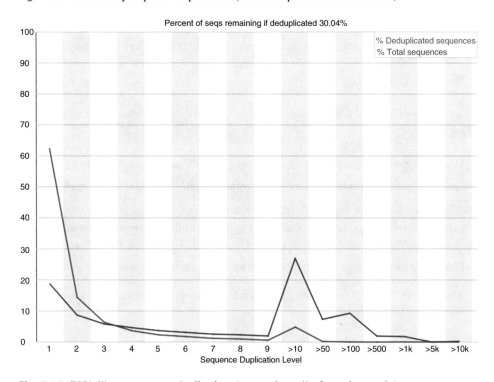

Fig. 7.16 RNA library sequence duplication. (source: https://rtsf.natsci.msu.edu)

7.3.1.11 Overrepresented Sequences

In a random library it is expected that most sequences occur only once in the final set. This module lists all sequences which appear more often than expected. Finding that a certain sequence is overrepresented either means that it is highly biologically relevant, or that the library is contaminated. For each detected overrepresented sequence, the program looks for matches in a database of common contaminants and will report the best hit it finds. Often adapter sequences are detected as overrepresented reads. This can occur if you use a long-read length and some of the library inserts are shorter than the read length resulting in read-through to the adapter at the 3' end of the read. How to remove adapter sequences is described in Sect. 7.3.2.

7.3.2 Preprocessing of NGS Data-Adapter Clipping

If you detected an adapter contamination in your *FastQC* output file, it is highly recommended to remove the adapter sequences from your sequences in the *fastq* file. Therefore, the package *cutadapt* can be used, which we have already installed by miniconda3. First you have to find the correct adapter sequence from the "Illumina Costumer Sequence Letter" (https://support.illumina.com/content/dam/illumina-support/documents/documentation/chemistry_documentation/experiment-design/illumina-adapter-sequences-1000000002694-11.pdf).

The basic usage of cutadapt is:

```
cutadapt -a ADAPTER [options] [-o output.fastq] input.fastq
```

For paired-end reads:

```
cutadapt -a ADAPT1 -A ADAPT2 [options] -o out1.fastq -p out2.fastq in1.fastq
in2.fastq
```

Moreover, you can also perform quality trimming with sequences depicting a bad phred score. This can be done with the additional option `cutadapt -` to trim low-quality bases from 5' and/or 3' ends of each read before adapter removal. Applied to both reads if data is paired. If one value is given, only the 3' end is trimmed. If two comma-separated cutoffs are given, the 5' end is trimmed with the first cutoff, the 3' end with the second.

Find some more options in terms of the command cutadapt by `cutadapt --help`.

Another tool for adapter trimming or removal of low-quality bases is `flexbar`.

Take Home Message
- The different sequence related formats include different information about the sequence.
- SRA is the file format in which all NCBI SRA content is provided.
- NGS sequence text files should be stored compressed to save up hard drive space.
- Quality Check of your raw reads is an essential step before further analysis.
- If any adapter contamination or low-quality bases at the end of your obtained sequencing reads are detected by the Quality Check tool, they should be removed.

Further Reading

- Sequence Read Archive Submissions Staff. Using the SRA Toolkit to convert *.sra files into other formats. In: SRA Knowledge Base [Internet]. Bethesda (MD): National Center for Biotechnology Information (US); 2011-. Available from: https://www.ncbi. nlm.nih.gov/books/NBK158900/

Answers to Review Questions:
Answer to Question 1: whole genome: [# of sequenced bases] / [size of genome]

$$\frac{300M \ x \ (2 \ x \ 100)}{3.2 \ GB} = \frac{45 \ x \ 10^9}{3.2 \ x \ 10^9} = 14.0625 \ (\text{coverage})$$

Answer to Question 2: Homo sapiens, transcriptome, Illumina HiSeq2500, 51, paired-end.

Answer to Question 3: 2 19 28 35 66 35 35 35 39 37 39 37 37 39 40 38 39 40 40 40

Answer to Question 4: To create a reference index to be able to perform sequence alignment.

Acknowledgements We thank Patricia Basurto and Carolina Castañeda of the International Laboratory for Human Genome Research (National Autonomous University of Mexico, Juriquilla campus) for reviewing this chapter.

References

1. Sims D, Sudbery I, Ilott NE, Heger A, Ponting CP. Sequencing depth and coverage: key considerations in genomic analyses. Nat Rev Genetics. 2014;15(2):121–32.
2. Cock PJ, Fields CJ, Goto N, Heuer ML, Rice PM. The Sanger FASTQ file format for sequences with quality scores, and the Solexa/Illumina FASTQ variants. Nucleic Acids Res. 2010;38 (6):1767–71.
3. Li H, Handsaker B, Wysoker A, Fennell T, Ruan J, Homer N, et al. The sequence alignment/map format and SAMtools. Bioinformatics. 2009;25(16):2078–9.

4. Robinson JT, Thorvaldsdottir H, Wenger AM, Zehir A, Mesirov JP. Variant review with the integrative genomics viewer. Cancer Res. 2017;77(21):e31–e4.

5. Robinson JT, Thorvaldsdottir H, Winckler W, Guttman M, Lander ES, Getz G, et al. Integrative genomics viewer. Nat Biotechnol. 2011;29(1):24–6.

6. Thorvaldsdottir H, Robinson JT, Mesirov JP. Integrative Genomics Viewer (IGV): high-performance genomics data visualization and exploration. Brief Bioinform. 2013;14(2):178–92.

7. Quinlan AR. BEDTools: the Swiss-army tool for genome feature analysis. Curr Protoc Bioinformatics. 2014;47:11–2. 1–34

8. Quinlan AR, Hall IM. BEDTools: a flexible suite of utilities for comparing genomic features. Bioinformatics. 2010;26(6):841–2.

9. Gericke A, Munson M, Ross AH. Regulation of the PTEN phosphatase. Gene. 2006;374:1–9.

10. Danecek P, Auton A, Abecasis G, Albers CA, Banks E, DePristo MA, et al. The variant call format and VCFtools. Bioinformatics. 2011;27(15):2156–8.

11. Leinonen R, Sugawara H, Shumway M. International nucleotide sequence database C. The sequence read archive. Nucleic Acids Res. 2011;39(Database issue):D19–21.

12. Trivedi UH, Cezard T, Bridgett S, Montazam A, Nichols J, Blaxter M, et al. Quality control of next-generation sequencing data without a reference. Front Genet. 2014;5:111.

Reference Genome

8

Melanie Kappelmann-Fenzl

Contents

What You Will Learn in This Chapter
This chapter describes the relevance of the reference genome for the analysis of Next
Generation Sequencing (NGS) data and how the respective reference genome can be
created. You will learn which databases provide the files for the creation of a
Reference Genome Index and which criteria you have to consider when choosing
the database and the respective files. Depending on the chosen alignment tool to be
used for further analyses, a Reference Genome Index must also be created with the
same tool. The corresponding code is shown in detail using the alignment software
tools *STAR* and *Bowtie2*.

M. Kappelmann-Fenzl (✉)
Deggendorf Institute of Technology, Deggendorf, Germany

Institute of Biochemistry (Emil-Fischer Center), Friedrich-Alexander University Erlangen-Nürnberg,
Erlangen, Germany
e-mail: melanie.kappelmann-fenzl@th-deg.de

© Springer Nature Switzerland AG 2021
M. Kappelmann-Fenzl (ed.), *Next Generation Sequencing and Data Analysis*, Learning
Materials in Biosciences, https://doi.org/10.1007/978-3-030-62490-3_8

8.1   Introduction

Depending on the samples sequenced (human, mouse, etc.) you need to generate a Genome Index of your reference genome before you are able to align your sequencing reads. Therefore, usually the comprehensive gene annotation on the primary assembly (chromosomes and scaffolds) sequence regions (PRI; .gtf file) and the nucleotide sequence (PRI, FASTA file) of the genome release of interest (e.g., GRCh38) are downloaded. Genome sequence and annotation files can be downloaded from various freely accessible databases as listed below:

- GENCODE: https://www.gencodegenes.org
- UCSC Genome Browser: https://hgdownload.soe.ucsc.edu/downloads.html
- Ensembl: https://www.ensembl.org/info/data/ftp/index.html
- NCBI RefSeq: https://www.ncbi.nlm.nih.gov/refseq/

Once the Genome sequence and annotation files have been downloaded, a Genome Index should be created. Each Genome Index has to be created by the software tool you are using for sequence alignment. In this chapter, we focus on the *STAR* and *Bowtie2* alignment tools.

8.2   Generate Genome Index via STAR

A key limitation with *STAR* [1] is its requirement for large memory space. *STAR* requires at least 30 GB to align to the human or mouse genomes. In order to generate the Genome Index with *STAR*; first, create a directory for the index (e.g., GenomeIndices/Star/ GRCh38_index). Then, copy the genome FASTA and Gene Transfer Format (GTF) files into this directory.

Example:

```
STAR -runMode genomeGenerate -runThreadN 23 -genomeDir /path/to/
GenomeIndices/Star/GRCh38_index/ -genomeFastaFiles /path/to/
GenomeIndices/Star/GRCh38_index/GRCh38.primary_assembly.genome.fa -
sjdbGTFfile /path/to/GenomeIndices/Star/GRCh38_index/gencode.v28.
primary_assembly.annotation.gtf -sjdbOverhang 99
```

```
Description of parameters:
--runMode                genomeGenerate
--runThreadN             Number of Threads
--genomeDir              /path/to/genomeDir
--genomeFastaFiles       /path/to/genome/fasta1 /path/to/genome/fasta2 ...
--sjdbGTFfile            /path/to/annotations.gtf
--sjdbOverhang           Read Length -1
```

For a more detailed explanation of the different options of *STAR* to build a Genome Index read the *STAR* Manual on GitHub:

https://github.com/alexdobin/STAR/blob/master/doc/STARmanual.pdf, or type `STAR -h`

in the terminal. After generating the Genome Index some more files with information about exonic gene sizes or chromosome sizes can be created using Bioconductor packages in R or the command line tool.

The detailed process on how to create these additional files is depicted below.

Create exonic gene sizes in R [2]

```
source("https://bioconductor.org/biocLite.R")
biocLite("GenomicFeatures")
setwd("/path/to/GenomeIndices/GRCh38")
library(GenomicFeatures)
txdb<-makeTxDbFromGFF("gencode.v24.primary_assembly.annotation.gtf",
format="gtf")
# then collect the exons per gene id
exons.list.per.gene <- exonsBy(txdb,by="gene")
# then for each gene, reduce all the exons to a set of non overlapping exons,
calculate their lengths (widths) and sum then
exonic.gene.sizes <- lapply(exons.list.per.gene,function(x)sum(width
(reduce(x))))
table <- t(exonic.gene.sizes)
write.table(t(table), file = "gencode.v24.primary_assembly.exonic.gene.
sizes.txt", sep = "  ", col.names=N
```

Extract geneID and gene symbol and gene_type from gtf annotation file

```
setwd("/path/to/GenomeIndices/GRCh38")
gtf.file = "gencode.v24.primary_assembly.annotation.gtf"
gtf.gr = rtracklayer::import(gtf.file) # creates a GRanges object
gtf.df = as.data.frame(gtf.gr)
genes = unique(gtf.df[,c("gene_id","gene_name","gene_type")])
library(data.table)
fwrite(genes, file="geneIDs_shortannotation.gencodeV24.txt", sep="\t")
```

Create the chromosome-size (command line):

```
samtools faidx /path/to/GenomeIndices/GRCh38/GRCh38.primary_assembly.
genome.fa cut -f1,2 /path/to/GenomeIndices/GRCh38/GRCh38.
primary_assembly.genome.fa.fai > GRCh38.chromosome.sizes
```

All the generated files should be stored in the *GenomeIndices/Star/*directory, or the name you have chosen.

8.3   Generate Genome Index via *Bowtie2*

Bowtie2 can also be used to generate Genome Index files (do not confuse *Bowtie2* indexing with *Bowtie* indexing as they are different). A more detailed description of *Bowtie* and *Bowtie2* can be found in Chap. 9. First, download FASTA files for the unmasked genome (i.e., hg38.fa.gz from http://hgdownload.cse.ucsc.edu/downloads.html) of interest if you have not already. Do NOT use masked sequences.

From the directory containing the *genome.fa* file, run the `bowtie2-build` command. The default options usually work well for most genomes. For example, for hg38:

```
bowtie2-build –threads 23 /path/to/GenomeIndices/bowtie2/GRCh38/GRCh38.
primary_assembly.genome.fa /path/to/GenomeIndices/bowtie2/GRCh38
```

This command will create 6 files with a **.bt2* file extension in your *Bowtie2* index directory. These will then be used by *Bowtie2* to map your sequencing data to the reference genome.

Take Home Message
- Generating a genome index is a time-consuming process, but you only need to do this once per reference genome.
- Organism and version of a reference genome are very important when mapping sequencing reads.
- To create an index of a reference genome you need the nucleotide sequence (FASTA) and the corresponding annotation file (GTF/GFF).
- The most common databases for reference genome download are: GENCODE, UCSC, Ensembl, and NCBI.
- Each Reference Genome Index must be created by the same software tool you want to use for alignment.

Acknowledgements We are grateful to Dr. Richa Barthi (Bioinformatician at TUM Campus Straubing, Germany) for critically reading this text. We thank for correcting our mistakes and suggesting relevant improvements to the original manuscript.

References

1. Dobin A, Gingeras TR. Optimizing RNA-Seq mapping with STAR. Methods Mol Biol. 2016;1415:245–62.
2. Lawrence M, Huber W, Pages H, Aboyoun P, Carlson M, Gentleman R, et al. Software for computing and annotating genomic ranges. PLoS Comput Biol. 2013;9(8):e1003118.

Alignment

9

Melanie Kappelmann-Fenzl

Contents

What You Will Learn in This Chapter
The purposes of the alignment process are to measure distances/similarities between strings and thus to locate origins of Next Generation Sequencing (NGS) reads in a reference genome. Alignment algorithms like BLAST that can be used to search for the location of a single or a small number of sequences in a certain genome are not

(continued)

M. Kappelmann-Fenzl (✉)
Deggendorf Institute of Technology, Deggendorf, Germany

Institute of Biochemistry (Emil-Fischer Center), Friedrich-Alexander University Erlangen-Nürnberg, Erlangen, Germany
e-mail: melanie.kappelmann-fenzl@th-deg.de

© Springer Nature Switzerland AG 2021
M. Kappelmann-Fenzl (ed.), *Next Generation Sequencing and Data Analysis*, Learning Materials in Biosciences, https://doi.org/10.1007/978-3-030-62490-3_9

suitable to align millions of NGS reads. This led to the development of advanced algorithms that can meet this task, allow distinguishing polymorphisms from mutations and sequencing errors from true sequence deviations. For a basic understanding, the differences between global and local alignment and the underlying algorithms are described in a simplified way in this chapter, as well as the main difference between BLAST and NGS alignment is described in a simplified way in this chapter. Moreover, different alignment tools and their basic usage are presented, which enables the reader to perform and understand alignment processes of sequencing reads to any genome using the respective commands.

9.1   Introduction

Sequence Alignment is a crucial step of the downstream analysis of NGS data, where millions of sequenced DNA fragments (reads) have to be aligned with a selected reference sequence within a reasonable time. However, the problem here is to find the correct position in the reference genome from where the read originates. Due to the repetitive regions of the genome and the limited length of the reads ranging from 50 to 150 bp, it often happens that shorter reads can map at several locations in the genome. On the other hand, a certain degree of flexibility for differences to the reference genome must be allowed during alignment in order to identify point mutations and other genetic changes.

Due to the massive amount of data generated during NGS analyses, all alignment algorithms use additional data structures (indices) that allow fast access and matching of sequence data. These indices are generated either over all generated reads or over the entire reference genome, depending on the used algorithm. Algorithms from computer science like hash tables or methods from data compression like suffix arrays are popularly implemented in the alignment tools. With the help of these algorithms, it is possible, for example, to compare over 100 GB of sequence data from NGS analyses with the human reference genome in just a few hours. With the help of high parallelization of computing capacity (CPUs), it is possible to reduce this time significantly. Thus, even large amounts of sequencing data from whole-exome or whole-genome sequencing can be efficiently processed.

9.2   Alignment Definition

The next step is the alignment of your sequencing reads to a reference genome or transcriptome. The main problem you have to keep in mind is that the human genome is really big and it is complex too. Sequencers are able to produce billions of reads per run and are prone to errors. Thus, an accurate alignment is a time-consuming process.

Fig. 9.1 Alignment definition. Sequence alignment is a way of arranging the sequences of DNA, RNA, or protein to identify regions of similarity. The basic principle is comparable to a puzzle (left). An optimal alignment means, an alignment with minimal errors like deletions, insertions, or mismatches—no error is defined as a match (right)

Sequence databases like GenBank (http://www.ncbi.nlm.nih.gov/genbank/) grew rapidly in the 1980s, and thus performing a full dynamic programming comparison of any query sequence to every known sequence soon became computationally very costly. Consequently, the alignment of a query sequence against a database motivated the development of a heuristic algorithm [1], which was implemented in the FASTA program suite [2]. The basic principle of this algorithm is to exclude large parts of the database from the expensive dynamic programming comparison by quickly identifying candidate sequences that share short sections (k-tuples) of very similar sequence with the query. FASTA was then followed by the BLAST program [3], with additional speed advantages and a new feature, which estimates the statistical likelihood that each matching sequence had been found by chance. BLAST is still one of the most used search program for biological sequence databases [4]. With the introduction of ultra-high-throughput sequencing technologies in 2007, other alignment challenges emerged. This chapter describes these efforts and the current state of the art in NGS alignment algorithms. Computational biologists have developed more than 70 read mapping to date [5]. A full list of sequence alignment software tools can be found at https://en.wikipedia.org/wiki/List_of_sequence_alignment_software#Short-Read_Sequence_Alignment. Actually, describing all of these tools is beyond the scope of this chapter, however main algorithmic strategies of these tools are depicted below.

Sequence alignment (Fig. 9.1) is widely used in molecular biology to find similar DNA, RNA, or protein sequences. These algorithms generally fall into two categories: global (Needleman–Wunsch), which aligns the entire sequence, and local (Smith–Waterman), which only look for highly similar subsequences.

9.2.1 Global Alignment (Needleman–Wunsch Algorithm)

Statistically the space for possible solutions is huge; however, we are interested in optimal alignments with minimal errors like indels or mismatches. The so-called unit edit distance (edist) is the number of mismatches, insertions, and deletions in an optimal sequence alignment. The main aim is to minimize the edist by tabulating partial solutions in a (m

$$E(i,j)= \min \begin{cases} E(i-1,j)+1 & \text{Deletion} \\ E(i,j-1)+1 & \text{Insertion} \\ E(i-1,j-1)+1 & \text{Substitution} \end{cases} \qquad E(i,j)= \max \begin{cases} E(i-1,j)-1 & \text{Deletion} \\ E(i,j-1)-1 & \text{Insertion} \\ E(i-1,j-1)-1 & \text{Substitution} \end{cases}$$

Fig. 9.2 Needleman–Wunsch. Optimization of distance (left) and optimization of similarity (right)

Needleman-Wunsch		
	Optimization of distance	Optimization of similarity
Match	0	0
Mismatch	1	-1
Gap	1	-1
Diagonal jump	Match/ Mismatch	
Top/ Down jump	Deletion	
Left/ Right jump	Insertion	

	E	i	G	C	A	C	T
j	0	1	2	3	4	5
T	1	1	2	3	4	4
G	2	1	2	3	4	5
A	3	2	2	2	3	4
T	4	3	3	3	3	3
A	5	4	4	3	4	4
T	6	5	5	4	4	4

```
T   G   A   T   A   T
    *   |   |   |   *
_   G   C   A   C   T
```

Fig. 9.3 Needleman–Wunsch Algorithm and the resulting Scoring Matrix (E). Matches are defined as 0, Mismatches and Gaps as 1/−1. The *edist* is marked in red [4]. A possible traceback is depicted by blue arrows and the corresponding alignment at the bottom right. Diagonal jumps within the scoring Matrix can be interpreted as Matches or Mismatches, Top or Down jumps as Deletions, and Left or Right jumps as Insertions

+1) x (n+1) matrix. Under the assumption that both input sequences a and b stem from the same origin, a global alignment tries to identify matching parts and the changes needed to transfer one sequence into the other. The changes are scored and an optimal set of changes is identified, which defines an alignment. The dynamic programming approach tabularizes optimal subsolutions in matrix E, where an entry E (i,j) represents the best score for aligning the prefixes $a_{1..i}$ with $b_{1..j}$ (Fig. 9.2).

Scoring Matrix using Needleman–Wunsch algorithm [6] and the corresponding traceback Matrix lead to the identification of the best alignment. One possible alignment result of our example and the related traceback are illustrated in Fig. 9.3.

9.2.2 Local Alignment (Smith–Waterman Algorithm)

Local alignment performed by the Smith–Waterman algorithm [7] aims to determine similarities between two nucleic acid or protein sequences. The main difference to the global alignment is that negative scoring matrix cells are set to zero (Fig. 9.5).

For a better understanding of local/sub-regions alignment imagine you have a little Toy genome (16 bp): CATGGTCATTGGTTCC.

Local alignment is a hash-based algorithm with two major approaches: hashing the reference and the Burrows–Wheeler transform [8, 9]. The first step is to hash/index the genome (forward strand only) resulting in a hash/k-mer index of your Toy genome:

k=3	K-mer/Hash	Positions
	CAT	1,7
	ATG	2
	TGG	3,10
	GGT	4,11
	GTC	5
	TCA	6
	ATT	8
	TTG	9
	GTT	12
	TTC	13
	TCC	14

Now, you want to align a Toy sequencing read (TGGTCA) to this indexed Toy genome. The k-mer index can be used to quickly find candidate alignment locations in the reference genome. For example, the k-mer TGG is assigned to Positions 3 and 10 and the k-mer TCA to position 6. Thus, Burrows–Wheeler transform is just another way of doing exact matches on hashes and check against genome and calculate a score.

This approach tries to identify the most similar subsequences that maximize the scoring of their matching parts and the changes needed to transfer one subsequence into the other. The dynamic programming approach tabularizes optimal subsolutions in matrix E (Fig. 9.4), where an entry $E_{i,j}$ represents the maximal similarity score for any local alignment of the (sub)prefixes $a_{x..i}$ with $b_{y..j}$, where x,y>0 are so far unknown and have to be identified via traceback (Fig. 9.5). Note: consecutive gap (Indels) scoring is done linearly.

Alignment to a reference genome can be performed with single- or paired-end sequencing reads, depending on your experiment and library preparation. Paired-end sequencing is recommended for RNA-Seq experiments.

Furthermore, we differ between two types of aligners:

- Splice unaware
- Splice aware

$$E\,(i,j)=\max \begin{cases} E(i\text{-}1,j)\text{-}1 & \text{Deletion} \\ E(i,j\text{-}1)\text{-}1 & \text{Insertion} \\ E(i\text{-}1,j\text{-}1)\text{-}1 & \text{Substitution} \end{cases}$$

Fig. 9.4 Smith–Waterman. Optimization of similarity

	Smith-Waterman
Match	2
Mismatch	-1
Gap	-1
Diagonal jump	Match/ Mismatch
Top/ Down jump	Deletion
Left/ Right jump	Insertion

E	i	G	C	A	C	T
j	0	0	0	0	0	0
T	0	0	0	0	0	2
G	0	2	1	0	0	1
A	0	1	1	3	2	1
T	0	0	0	2	2	4
A	0	0	0	2	1	3
T	0	0	0	1	1	3

```
G   _   A   _   T
*       *       *
G   C   A   C   T
```

Fig. 9.5 Smith–Waterman Algorithm and the resulting Scoring Matrix (E). Matches are defined as 2, Mismatches and Gaps as −1. The traceback is depicted by blue arrows and the corresponding alignment at the bottom right. Diagonal jumps within the scoring Matrix can be interpreted as Matches or Mismatches, Top or Down jumps as Deletions, and Left or Right jumps as Insertions

Table 9.1 Splice-aware and splice-unaware alignment tools

Alignment tool	Splice-aware	Link
STAR	Yes	https://github.com/alexdobin/STAR
Bowtie	No	http://bowtie-bio.sourceforge.net/index.shtml
Bowtie2	Yes	http://bowtie-bio.sourceforge.net/bowtie2/index.shtml
TopHat/TopHat2	Yes	http://ccb.jhu.edu/software/tophat/index.shtml
BWA-MEM	Yes	http://bio-bwa.sourceforge.net/
BWA-SW	No	
BWA-backtrack	No	
Hisat2	Yes	https://ccb.jhu.edu/software/hisat2/manual.shtml
Segemehl	Yes	https://www.bioinf.uni-leipzig.de/Software/segemehl/

Splice-unaware aligners are able to align continuous reads to a reference genome, but are not aware of exon/intron junctions. Hence, in RNA-sequencing, splice-unaware aligners are no proper tool to analyze the expression of known genes, or align reads to the transcriptome. Splice-aware aligners map reads over exon/intron junctions and are therefore used for discovering new splice forms, along with the analysis of gene expression levels (Table 9.1).

In this context the most common alignment tools are explained in the following section.

9.2.3 Alignment Tools

9.2.3.1 STAR

Spliced Transcripts Alignment to a Reference (STAR) is a standalone software that uses sequential maximum mappable seed search followed by seed clustering and stitching to align RNA-Seq reads. It is able to detect canonical junctions, non-canonical splices, and chimeric transcripts.

The main advantages of STAR are its high speed, exactness, and efficiency. STAR is implemented as a standalone C++ code and is freely available on GitHub (https://github.com/alexdobin/STAR/releases) [10].

In terms of mapping multiple samples, you can parallelize your mapping command. First, create a .txt file containing your file names:

```
parallel -j 1 echo {1} ::: Sample1 Sample2 Sample3 ::: > /path/to/SampleNames.txt
```

You can use the generated *SampleNames.txt* file to combine the commands for mapping and sorting and run it on various samples:

```
cat /path/to/SampleNames.txt | parallel -j 1 "add your mapping commands here
and write "{}" whenever the sample name appears within your commands"
```

Example- Mapping command via STAR (RNA-Seq, paired-end):

```
cat /path/to/SampleNames.txt | parallel -j 1 "mkdir /path/to/mapping_hg38/
{} ; cd /path/to/mapping_hg38/{} ; STAR –runThreadN 15 –genomeDir /path/to/
GenomeIndices/Star/GRCh38_index_100 –readFilesIn /path/to/rawData/{}_r1.
fastq /path/to/rawData/{}_r2.fastq –outFilterType BySJout –outFilter
MultimapNmax 20 –outFilterIntronMotifs RemoveNoncanonicalUnannotated –
outReadsUnmapped Fastq –alignSJoverhangMin 8 –alignSJDBoverhangMin 1 –
alignMatesGapMax 1000000 –alignIntronMax 1000000 –outSAMtype BAM
SortedByCoordinate –quantMode GeneCounts –outWigType bedGraph –
outWigStrand Stranded"
```

The mapping job can be checked in the *Log.progress.out* file in the run directory. This file is updated every minute and shows the number of reads that have been processed, and various mapping statistics. This is useful for initial quality control during the mapping job.

Log.final.out contains the summary mapping statistics of the run.

In the next step, the bedgraph files are sorted and converted to bigwig files (bedGraphToBigWig).

Example:

```
cat /path/to/SampleNames.txt | parallel -j 1 "LC_COLLATE=C sort -k1,1 -k2,2n
/path/to/ /mapping_hg38/{}/Signal.Unique.str1.out.bg > /path/to/
mapping_hg38/{}/Signal.Unique.str1.out.sorted.bg ; bedGraphToBigWig /
path/to/mapping_hg38/{}/Signal.Unique.str1.out.sorted.bg /path/to/
GenomeIndices/Star/GRCh38_index_100/GRCh38.chromosome.sizes /path/to/
bigWig/{}.Signal.Unique.str1.out.bigWig;
```

9.2.3.2 Bowtie

Bowtie is an ultrafast and memory-efficient short read alignment tool to large reference genomes indexed with a Burrows–Wheeler index. It is typically used for aligning DNA sequences as it is a splice-unaware tool. Because of this feature this tool is often used in microbiome alignment (http://bowtie-bio.sourceforge.net/index.shtml) [11, 12].

```
bowtie [options]* -x <ebwt> {-1 <m1> -2 <m2> | -12 <r> | -interleaved <i> | <s>}
[<hit>]
```

```
Explanation
-x                      path to Bowtie index
-q                      query input files are FASTQ
-1 <m1>                 fastq input files (first mate)
-2 <m2>                 fastq input files (second mate)
-c <s>                  unpaired reads
--12 <r>                mixed sequencing reads (unpaired and paired-end)
--interleaved <i>       interleaved paired-end FASTQ files
```

9.2.3.3 Bowtie2

Bowtie2, as well as *Bowtie*, is an ultrafast and memory-efficient tool, but more suitable for aligning sequencing reads of about 50 up to 100s or 1,000s of characters to relatively long reference sequences (e.g., mammalian genomes) indexed with an Ferragina–Manzini (Fm) index. *Bowtie2* supports gapped, local, and paired-end alignment modes (https://github.com/BenLangmead/bowtie2) [13].

```
#Single-end sequencing reads
bowtie2 -x /path/to/GenomeIndices/bowtie2/GRCh38 -U *.fq
#Paired-end sequencing reads
bowtie2 -x /path/to/GenomeIndices/bowtie2/GRCh38 -1 *.1_fq -2 *.2_fq
```

Explanation	
-x	path to *Bowtie2* index
-U <r>	unpaired reads
-1 <m1>	fastq input files (first mate)
-2 <m2>	fastq input files (second mate)

9.2.3.4 TopHat/TopHat2

TopHat aligns RNA-Seq reads to genomes by first using the short-read aligner *Bowtie*, and then by mapping to a reference genome to discover RNA splice sites *de novo*. RNA-Seq reads are mapped against the whole reference genome, and those reads that do not map are set aside. TopHat is often paired with the software Cufflinks for a full analysis of sequencing data (https://github.com/dnanexus/tophat_cufflinks_RNA-Seq/tree/master/tophat2) [14].

A detailed description of the usage of TopHat can be found in the TopHat manual (http://ccb.jhu.edu/software/tophat/manual.shtml).

```
tophat [options]* <genome_index_base> <reads1_1[,...,readsN_1]>
[reads1_2,...readsN_2]
```

Explanation	
<genome_index_base>	path to *TopHat* index
<reads1_1>	fastq input files (unpaired or first mate if paired-end)
<reads1_2>	fastq input files (second mate if paired-end)

9.2.3.5 Burrow–Wheeler Aligner (BWA)

BWA is a splice-unaware software package for mapping low-divergent sequences against a large reference genome, such as the human genome. It consists of three algorithms: BWA-backtrack, BWA-SW, and BWA-MEM. The first algorithm is designed for Illumina sequence reads up to 100bp. BWA-MEM and BWA-SW share similar features such as long-read support and split alignment, but BWA-MEM (maximal exact matches), which is the latest, is generally recommended for high-quality queries as it is faster and more accurate (https://github.com/lh3/bwa) [8, 9]. The splice-unaware alignment algorithms are recommended for species like bacteria.

A detailed description of the usage of BWA can be found in the BWA manual (http://bio-bwa.sourceforge.net/bwa.shtml).

```
bwa mem /path/to/GenomeIndices/BWA/GRCh38/*.fa reads.fq [mates.fq]
```

```
Description of parameters:
*.fa                     path to BWA index file
reads.fq                 fastq input files (unpaired or first mate if
                         paired-end)
mates.fq                 fastq input files (second mate if paired-end)
```

9.2.3.6 HISAT2

HISAT (and its newer version HISAT2) is the next generation of spliced aligner from the same group that has developed TopHat. HISAT uses an indexing scheme based on the Burrows–Wheeler transform and the Ferragina–Manzini (Fm) index, employing two types of indices for alignment: a whole-genome Fm index to anchor each alignment and numerous local Fm indexes for very rapid extensions of these alignments (https://github.com/DaehwanKimLab/hisat) [15].

HISAT most interesting features include its high speed and its low memory requirement. HISAT is an open-source software freely available. A detailed description of the usage of HISAT can be found in the HISAT manual (https://ccb.jhu.edu/software/hisat2/manual.shtml).

```
hisat2 [options]* -x <hisat2-idx> {-1 <m1> -2 <m2> } | -U <r>
```

```
Description of parameters:
-x                       path to Hisat2 index
-U <r>                   unpaired reads
-1 <m1>                  fastq input files (first mate)
-2 <m2>                  fastq input files (second mate)
```

Take Home Message
- Sequence alignment is the process of comparing and detecting distances/similarities between biological sequences.
- Dynamic programming technique can be applied to global alignments by using methods such as global and local alignment algorithms.
- The value that measures the degree of sequence similarity is called the *alignment score*.

(continued)

- Sequence alignment includes calculating the so-called *edit distance,* which generally corresponds to the minimal number of substitutions, insertions, and deletions needed to turn one sequence into another.
- The choice of a sequencing read alignment tool depends on your goals and the specific case.

Review Questions

1. Assume you are performing a mapping with STAR. After the process, where do you find the information how many reads have mapped only at one position of the reference genome?
2. Which of the following does not describe local alignment?
   A. A local alignment aligns a substring of the query sequence to a substring of the target sequence.
   B. A local alignment is defined by maximizing the alignment score, so that deleting a column from either end would reduce the score, and adding further columns at either end would also reduce the score.
   C. Local alignments have terminal gap.
   D. The substrings to be examined may be all of one or both sequences; if all of both are included, then the local alignment is also global.
3. Which of the following does not describe BLAST?
   A. It stands for Basic Local Alignment Search Tool.
   B. It uses word matching like FASTA.
   C. It is one of the tools of the NCBI.
   D. Even if no words are similar, there is an alignment to be considered.
4. Which of the following does not describe dynamic programming?
   A. The approach compares every pair of characters in the two sequences and generates an alignment, which is the best or optimal.
   B. Global alignment algorithm is based on this method.
   C. Local alignment algorithm is based on this method.
   D. The method can be useful in aligning protein sequences to protein sequences only.
5. Which of the following is not a disadvantage of Needleman–Wunsch algorithm?
   A. This method is comparatively slow.
   B. There is a need of intensive memory.
   C. This cannot be applied on genome sized sequences.
   D. This method can be applied to even large sized sequences.
6. Alignment algorithms, both global and local, are fundamentally similar and only differ in the optimization strategy used in aligning similar residues.
   A. True.
   B. False.

7. The function of the scoring matrix is to conduct one-to-one comparisons between all components in two sequences and record the optimal alignment results.
   A. True.
   B. False.

Answers to Review Questions

1. *Log.final.out* in the output directory of alignment file (.bam); 2. C; 3. D; 4. D; 5. D; 6. A; 7. A

Acknowledgements We are grateful to Dr. Richa Bharti (Bioinformatician at TUM Campus Straubing, Germany) and Dr. Philipp Torkler (Senior Bioinformatics Scientist, *Exosome Diagnostics, a Bio-Techne brand*, Munich, Germany) for critically reading this text. We thank for correcting our mistakes and suggesting relevant improvements to the original manuscript.

References

1. Wilbur WJ, Lipman DJ. Rapid similarity searches of nucleic acid and protein data banks. Proc Natl Acad Sci U S A. 1983;80(3):726–30.
2. Pearson WR, Lipman DJ. Improved tools for biological sequence comparison. Proc Natl Acad Sci U S A. 1988;85(8):2444–8.
3. Altschul SF, Gish W, Miller W, Myers EW, Lipman DJ. Basic local alignment search tool. J Mol Biol. 1990;215(3):403–10.
4. Canzar S, Salzberg SL. Short read mapping: an algorithmic tour. Proc IEEE Inst Electr Electron Eng. 2017;105(3):436–58.
5. Fonseca NA, Rung J, Brazma A, Marioni JC. Tools for mapping high-throughput sequencing data. Bioinformatics. 2012;28(24):3169–77.
6. Needleman SB, Wunsch CD. A general method applicable to the search for similarities in the amino acid sequence of two proteins. J Mol Biol. 1970;48(3):443–53.
7. Smith TF, Waterman MS. Identification of common molecular subsequences. J Mol Biol. 1981;147(1):195–7.
8. Li H, Durbin R. Fast and accurate short read alignment with Burrows-Wheeler transform. Bioinformatics. 2009;25(14):1754–60.
9. Li H, Durbin R. Fast and accurate long-read alignment with Burrows-Wheeler transform. Bioinformatics. 2010;26(5):589–95.
10. Dobin A, Gingeras TR. Optimizing RNA-seq mapping with STAR. Methods Mol Biol. 2016;1415:245–62.
11. Langmead B. Aligning short sequencing reads with Bowtie. Curr Protoc Bioinformatics. 2010;32:11–7.
12. Langmead B, Trapnell C, Pop M, Salzberg SL. Ultrafast and memory-efficient alignment of short DNA sequences to the human genome. Genome Biol. 2009;10(3):R25.
13. Langmead B, Salzberg SL. Fast gapped-read alignment with Bowtie 2. Nat Methods. 2012;9 (4):357–9.
14. Kim D, Pertea G, Trapnell C, Pimentel H, Kelley R, Salzberg SL. TopHat2: accurate alignment of transcriptomes in the presence of insertions, deletions and gene fusions. Genome Biol. 2013;14(4): R36.
15. Kim D, Langmead B, Salzberg SL. HISAT: a fast spliced aligner with low memory requirements. Nat Methods. 2015;12(4):357–60.

Identification of Genetic Variants and de novo Mutations Based on NGS

10

Patricia Basurto-Lozada, Carolina Castañeda-Garcia, Raúl Ossio, and Carla Daniela Robles-Espinoza

Contents

P. Basurto-Lozada · C. Castañeda-Garcia · R. Ossio
Laboratorio Internacional de Investigación sobre el Genoma Humano, Universidad Nacional
Autónoma de México, Campus Juriquilla, Santiago de Querétaro, México

C. D. Robles-Espinoza (✉)
Laboratorio Internacional de Investigación sobre el Genoma Humano, Universidad Nacional
Autónoma de México, Campus Juriquilla, Santiago de Querétaro, México

Experimental Cancer Genetics, Wellcome Sanger Institute, Cambridge, UK
e-mail: drobles@liigh.unam.mx

© Springer Nature Switzerland AG 2021
M. Kappelmann-Fenzl (ed.), *Next Generation Sequencing and Data Analysis*, Learning
Materials in Biosciences, https://doi.org/10.1007/978-3-030-62490-3_10

What You Will Learn in This Chapter
In this chapter, we will discuss an overview of the bioinformatic process for the identification of genetic variants and *de novo* mutations in data recovered from NGS applications. We will pinpoint critical steps, describe the theoretical basis of different variant calling algorithms, describe data formats, and review the different filtering criteria that can be undertaken to obtain a set of high-confidence mutations. We will also go over crucial issues to take into account when analyzing NGS data, such as tissue source or the choice of sequencing machine. We also discuss different methodologies for analyzing these variants depending on study context, considering population-wide and family-focused analyses. Finally, we also do an overview of available software for variant filtering and genetic data visualization.

10.1 Introduction: Quick Recap of a Sequencing Experiment Design

As we have seen throughout this book, NGS applications give researchers an all-access pass to the building information of all biological organisms. After establishing the biological question to be pursued and once the organism of interest has been sequenced, the first step is to align this information against a reference genome (see Chap. 8). This reference genome should be one that is as biologically close as it can be to the subject of interest—if there is no reference genome or the one available is not reliable, then a possible option is to attempt to build one (See Box: Genome assembly). After read mapping and alignment, and quality control, one or several variant callers will need to be run to identify the variants present in the query sequence. Finally, depending on the original aim, different post-processing and filtering steps may also need to be deployed to extract meaningful information out of the experiment.

10.2 How Are Novel Genetic Variants Identified?

The correct identification of variants depends on having accurately performed base calling, and read mapping and alignment previously. "Base calling" refers to the determination of the identity of a nucleotide from the fluorescence intensity information outputted by the sequencing instrument. Read mapping is the process of determining where a read originates from, using the reference genome, and read alignment is the process of finding the exact differences between the sequences. These topics have already been reviewed throughout this book (Chaps. 4, 8 and 9). Base calling and read alignment results rely on the sequencing instrument and algorithm used; therefore, it is important to state the confidence we have on the assignment of each base. This is expressed by standard Phred quality scores [1].

$$Q\text{Phred} = -10 \, \log_{10} P(\text{error})$$

This measurement is an intuitive number that tells us the probability of the base call or the alignment being wrong, and the higher Q is, the more confidence we have that there has not been an error. For example, if $Q = 20$, then that means there is a 1 in 100 chance of the call or alignment being wrong, whereas if it is 30 then there is a 1 in 1000 chance of a mistake. Subsequently, steps such as duplicate read marking and base call quality score recalibration can be performed (See Chap. 7, Sect. 7.2.3).

Review Question 1

What would be the value of Q for a variant that has a 1 in 3000 chance of being wrong?

After the previous steps have taken place, and an alignment file has been produced (usually in the BAM and CRAM file formats), the next step is to identify differences between the reference genome and the genome that has been sequenced. To this effect, there are different strategies that a researcher can use depending on their experiment, for example, for germline analyses they might use algorithms that assume that the organism of interest is diploid (or another, fixed ploidy) and for cancer genomes they may need to use more flexible programs due to the presence of polyploidy and aneuploidy. In this Chapter, we will focus on the former analyses, but the reader is referred to the publications on somatic variant callers in the "Further Reading" section below if they want to learn more.

When identifying variants, and particularly if a researcher is performing whole exome or genome sequencing, the main objective is to determine the genotype of the sample under study at each position of the genome. For each variant position there will be a reference (R) and an alternate (A) allele, the former refers to the sequence present in the reference genome. Therefore, it follows that in the case of diploid organisms, there will be three different possible genotypes: RR (homozygous reference), RA (heterozygous), and AA (homozygous alternative).

10.2.1 Naive Variant Calling

A naive approach to determining these genotypes from a pile of sequencing reads mapped to a site in the genome may be to count the number of reads with the reference and alternate alleles and to establish hard genotype thresholds; for example, if more than 75% of reads are R, then the genotype is called as RR; if these are less than 25%, then the genotype is called as AA; and anything in between is deemed RA. However, even if careful steps are taken to ensure that only high-quality bases and reads are counted in the calculation, this method is still prone to under-calling variants in low-coverage data, as the counts from

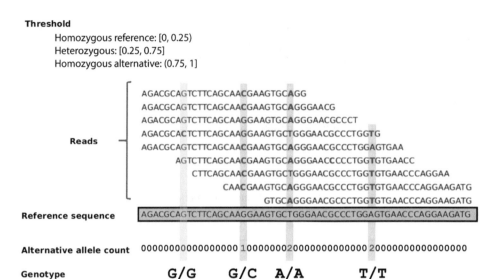

Fig. 10.1 Naive variant calling. In this method, reads are aligned to the reference sequence (green) and a threshold of the proportion of reads supporting each allele for calling genotypes is established (top). Then, at each position, the proportion of reads supporting the alternative allele is calculated and, based on the dosage of the alternative allele, a genotype is established. Yellow: a position where a variant is present but the proportion of alternative alleles does not reach the threshold ($1/6 < 0.25$). In light blue, positions where a variant has been called. This figure is based on one drawn by Petr Danecek for a teaching presentation

these reads will likely not reach the set threshold, relevant information can be ignored due to hard quality filters, and also it would not give any measure of confidence [2] (Fig. 10.1). Even though this type of method was used in early algorithms, it has been dropped in favor of other algorithms that are able to deal with errors and low-coverage data better.

10.2.2 Bayesian Variant Calling

The sequencing of two mixed molecules of DNA is a probabilistic event, as well as the occurrence of errors in previous steps of this process, and therefore, a more informative approach would include information about the prior probability of a variant occurring at that site and the amount of information supporting each of the potential genotypes. To address this need, most currently used variant callers implement a Bayesian probabilistic approach at their core in order to assign genotypes [3]. Examples of the most commonly used algorithms using this method are GATK HaplotypeCaller [4] and bcftools mpileup (Formerly known as Samtools mpileup) [5].

This probabilistic approach uses the widely known Bayes' Theorem, which, in this context, is able to express the posterior probability of a genotype given the sequencing data as the product of a prior genotype probability and the genotype likelihood, divided by a constant:

$$P(G|D) = \frac{P(D|G)\,P(G)}{P(D)}$$

where

- P(G|D) is the posterior probability of the genotype given the sequencing data
- P(D|G) is the genotype likelihood
- P(G) is the prior genotype probability
- P(D) is a factor to normalize the sum of all posterior probabilities to 1, it is constant throughout all possible genotypes

As this is the most commonly used method for variant calling, has been for a number of years and is unlikely to change, special attention should be given to it to understand its basics. For deciding what the genotype is at a particular site in an individual, a variant calling algorithm would calculate the posterior probability P(G|D) for each possible genotype, and pick the genotype with the highest one. As the sum of all posterior genotype probabilities must be equal to 1, the number of different possible genotypes that the algorithm (referred to as "genotype partitioning") considers is crucial. Different algorithms will calculate these differently, for example, some algorithms may only consider three genotype classes: homozygous reference, heterozygous reference, and all others, whereas others may consider all possible genotypes [6]. The choice of algorithm would depend on the original biological question: For example, if tumors are being analyzed, which can be aneuploid or polyploid, an algorithm that only considers three possible genotypes may be inadequate.

The prior genotype probability, P(G), can be calculated taking into account the results of previous sequencing projects. For example, if a researcher is sequencing humans, the prior probability of a genotype being found at a particular site could depend on previously reported allele frequencies at that site as well as the Hardy–Weinberg Equilibrium principle [2]. Information from linkage disequilibrium calculations can also be incorporated. Otherwise, if no information is available, then the prior probability of a variant occurring at a site may be set as a constant for all loci. Algorithms also differ in the way they calculate these priors, and the information they take into account. The denominator of the equation, P(D), remains constant throughout all genotypes being considered and serves to normalize all posterior probabilities so they sum up to 1. Therefore, it is equal to the sum of all numerators, $P(D) = \Sigma\, P(D|Gi)\, P(Gi)$, where Gi is the ith genotype being considered.

The last part of the equation is the genotype likelihood, P(D|G). This can be interpreted as the probability of obtaining the sequencing reads we have given a particular genotype. It can be calculated from the quality scores associated with each read at the site being considered, and then multiplying these across all existing reads, assuming all reads are independent [2]. For example, in what is perhaps the most commonly used variant calling algorithm, GATK HaplotypeCaller [4], a number of steps are followed to determine genotype likelihoods: First, regions of the genome where there is evidence of a variant are defined ("SNP calling"), then the sequencing reads are used to identify the most likely genotypes supported by these data, and each read is re-aligned to all these most likely haplotypes in order to obtain the likelihoods per haplotype and per variant given the read data. These are then input into Bayes' formula to identify the most likely genotype for a sample [4].

10.2.3 Heuristic Variant Calling

Other methods that do not rely on naive or Bayesian approaches have been developed; these methods rely on heuristic quantities to call a variant site, such as a minimum coverage and alignment quality thresholds, and stringent cut-offs like a minimum number of reads supporting a variant allele. If performing somatic variant calling, a statistical test such as a Fisher's exact comparing the number of reference and alternate alleles in the tumor and normal samples is then performed in order to determine the genotype at a site [7]. Parameters used for variant calling can be tuned, and generally this method will work well with high-coverage sequencing data, but may not achieve an optimal equilibrium between high specificity and high sensitivity at low to medium sequencing depths, or when searching for low-frequency variants in a population [8].

The following GATK commands depict an example workflow for calling variants in NGS data. The installation instruction is covered in Chap. 5.

First you can check all available GATK tools by typing `gatk-list`. If not already done, you also have to install all required software packages (http://gatkforums.broadinstitute.org/gatk/discussion/7098/howto-install-software-for-gatk-workshops) for GATK analyses workflows. Moreover, be sure to set the PATH in your *.bashrc* to your GATK executable PATH.

The GATK (v4) uses two files to access and safety check access to the reference files: a *.dict* dictionary of the contig names and sizes and a *.fai* fasta index file to allow efficient random access to the reference bases. You have to generate these files in order to be able to use a Fasta file as reference.

```
#Create .fai fasta index file:
samtools faidx /path/to/GenomeIndices/GRCh38_index_100/GRCh38.
primary_assembly.genome.fa
#Create .dict dictionary of the contig names and sizes file:
gatk CreateSequenceDictionary R=GRCh38.primary_assembly.genome.fa
O=GRCh38.primary_assembly.genome.dict
#Add read groups, sort, mark duplicates, and create index
```

```
gatk AddOrReplaceReadGroups I=/path/to/Aligned.sortedByCoord.out.sam
O=sampleName_rg_added_sorted.bam SO=coordinate RGID=id RGLB=library
RGPL=platform RGPU=machine RGSM=sample
gatk MarkDuplicates I=sampleName_rg_added_sorted.bam O=dedup_sampleName.
bam M=metrics.txt
gatk BuildBamIndex I=dedup_sampleName.bam
```

Next, we use a GATK tool called SplitNCigarReads developed specially for RNAseq, which splits reads into exon segments (getting rid of Ns but maintaining grouping information) and hard-clip any sequences overhanging into the intronic regions.

```
#Split'N'Trim and reassign mapping qualities
gatk SplitNCigarReads -R path/to/GenomeIndices /GRCh38_index_100/GRCh38.
primary_assembly.genome.fa -I dedup_sampleName.bam -o
split_dedup_sampleName.bam -rf ReassignOneMappingQuality -RMQF 255 -RMQT 60
-U ALLOW_N_CIGAR_READS
```

In this example we will use Mutect2 to perform variant calling, which identifies somatic SNVs and indels via local assembly of haplotypes.

```
#Variant calling
gatk Mutect2 -R /path/to/GenomeIndices/GRCh38_index_100/GRCh38.
primary_assembly.genome.fa -I split_sampleName.bam -o
CalledVariants_sampleName.out.vcf
#Variant Filtration
gatk VariantFiltration -R /path/to/GenomeIndices/GRCh38_index_100/GRCh38.
primary_assembly.genome.fa -V CalledVariants_sampleName.out.vcf -window
35 -cluster 3 -filterName FS -filter "FS > 30.0" -filterName QD -filter "QD <
2.0" -o FilteredCalledVariants_sampleName.out.vcf
```

For a more detailed description see https://github.com/gatk-workflows/gatk4-jupyter-notebook-tutorials/blob/master/notebooks/Day3-Somatic/1-somatic-mutect2-tutorial.ipynb.

10.2.4 Other Factors to Take into Account When Performing Variant Calling

As we have seen, errors can be introduced at every step of the variant calling process. On top of errors brought in during the base calling and read mapping and alignment steps, other factors that can influence data quality are the preparation and storage of samples prior to analysis. For example, it is known that samples stored as formalin-fixed paraffin embedded (FFPE) tissue will have a higher bias toward C>T mutations due to deamination events triggered by a long fixation time [9]. Although these events are detectable only at a small fraction of the reads aligning to a particular site, they can become important when analyzing a pool of genomes sequenced at low frequency or when studying tumor samples that could have subclonal mutations—furthermore confounded by the tendency of some of these tumor types toward having more real C>T mutations [10]. Another example comes from the observation that DNA oxidation can happen during the shearing step most NGS protocols have implemented, and that this results in artifactual C>A mutations [11]. Ancient DNA and ctDNA can also suffer from these problems [12, 13]. Therefore, a researcher needs to consider their sample origin and preparation protocol and undertake post-processing filtering steps accordingly.

10.2.5 How to Choose an Appropriate Algorithm for Variant Calling?

In addition to considering the variant calling method (*e.g.,* naive, probabilistic, or heuristic) that an algorithm implements, a researcher also needs to consider the types of genetic variants that they are interested in analyzing, perhaps having to run several programs at the same time to obtain a comprehensive picture of the genetic variation in their samples.

Genetic variants are usually classified into several groups according to their characteristics (Fig. 10.2):

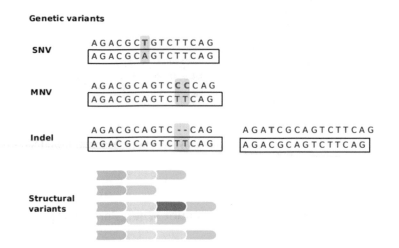

Fig. 10.2 Classes of genetic variants. Genetic variants ranging from a single base change, to the insertion or deletion of several bases can occur in a genome. Structural variants are more complex and encompass larger sections of a genome: At the top, a reference sequence, in the second row, a large deletion (blue region), in the third row, a large insertion (red section), in the fourth row, an inversion, and in the fifth row, a duplication. This figure is based on one drawn by Petr Danecek for a teaching presentation

- *SNVs (single nucleotide variants)*, also known as single base substitutions, are the simplest type of variation as they only involve the change of one base for another in a DNA sequence. These can be subcategorized into transitions (Ti) and transversions (Tv); the former are changes between two purines or between two pyrimidines, whereas the latter involve a change from a purine to a pyrimidine or vice versa. An example of a transition would be a G > A variant. If the SNV is common in a population (usually with an allele frequency > 1%), then it is referred to as a SNP (single nucleotide polymorphism). A common post-calling analysis involves looking at the Ti/Tv ratio, which can vary between 2 and 3 depending on the genomic region under analysis [14]. If this ratio is far from the expected, it may indicate a large proportion of false positive calls.
- *MNVs (multi-nucleotide variants)*, which are sequence variants that involve the consecutive change of two or more bases. An example would be one of the types of mutations caused by UV irradiation, CC>TT. Similarly to SNVs, there are some MNVs that are found at higher frequencies in the population, which are referred to as MNPs [15].
- *Indels (portmanteau of insertions and deletions)*, which involve the gain or loss of one or more bases in a sequence. Usually, what is referred to as indel tends to be only a few bases in length. An example of a deletion would be CTGGT > C and an insertion would be represented as T > TGGAT.
- *Structural variants*, which are genomic variations that involve larger segments of the genome. These can involve inversions, which is when a certain sequence in the genome gets reversed end to end, and copy number variants including amplifications, when a fraction of genome gets duplicated one or more times, and larger deletions, when large segments of the genome get lost. There is not a strict rule defining the number of base pairs that make the difference between an indel and a structural variant, but usually, a gain or loss of DNA would be called a structural variant if it involved more than one kilobase of sequence.

Most variant callers identify SNVs, but there are only some variant callers that will report indels or structural variation [3]. This is because usually the algorithms underlying the detection of these types of variants tend to be quite different: SNV, MNV, and short indel detection comprise the comparison of a pile of sequencing reads and their alignments to the reference genome (as has been discussed throughout Chap. 9), whereas larger indels and structural variant calling require calculating a distribution of insert sizes and detecting those read pairs that fall outside it, as well as the direction of alignment of both mate pairs [16].

It is also important to consider the type of sequencing that was used for the experiment. For example, whole genome sequencing and whole exome sequencing have different amounts of coverage, depth, and sequencing uniformity. Some variant callers such as MuTect2 and Strelka2 show better performance in sequencing with higher average sequencing depth and lower coverage [17].

10.3   Working with Variants

Variant calling usually outputs VCF files (See Sect. 7.2 File Formats) [18]. To recap, VCF files are plain text files that contain genotype information about all samples in a sequencing project. A VCF file is arranged like a matrix, with chromosome positions in rows and variant and sample information in columns. The sample information contains the genotype called by the algorithm along with a wealth of information such as (depending on the algorithm) genotype likelihoods and sequencing depth supporting each possible allele, among others. For each variant position, the file also contains information outputted by the variant caller such as the reference and alternate alleles, the Phred-based variant quality score, whether overlapping variants in other sequencing or genotype projects have been found, etc. Crucially, this file also contains a column called "FILTER," where information about whether further quality filters have been applied to the calls and which ones. We will review here some of the most common filters that researchers should consider applying to their data once it has already been called.

Review Question 2

How do you think a researcher can deal with the uncertainty about false negatives, *i.e.* sites where a variant has not been called? How can they be sure there is no variation there and it is not, let us say, a lack of sequence coverage?

10.4   Applying Post-variant Calling Filters

So far, we have seen a number of steps where a researcher must be careful to increase both the sensitivity and specificity of their set of calls in order to have an accurate view of the amount and types of sequencing variation present in their samples. However, there are also a number of post-calling filtering steps that should be applied in the majority of cases in order to further minimize the amount of false positive calls. Here we will review the main such filters, but the reader is referred to [18] if they wish to delve deeper into this topic.

Strand Bias Filter . Sometimes a phenomenon, referred to as strand bias, can be observed where one base is called only in reads in one direction, whereas it is absent in the reads in the other direction. This is evidently an error introduced during the preceding steps, and can be detected through a strand bias filter. This filter applies a Fisher's exact test comparing the number of forward and reverse reads with reference and alternate alleles, and if the P-value is sufficiently small as determined through a pre-chosen threshold, then the variant is deemed an artifact.

Variant Distance and End-Distance Bias Filters These filters were primarily developed to deal with RNA sequencing data when aligned to a reference genome [19]. Therefore, if a variant is mostly or only supported by differences in the last bases of each read, the call may be a false positive resulting from a portion of a read coming from a mRNA being aligned to an intron adjacent to a splice junction. For end-distance bias, a *t*-test is performed in order to determine whether the variants occur at a fixed distance from read ends, whereas for the variant distance bias tests whether or not variant bases occur at random positions in the aligned reads [19].

Indel and SNP Gap Filters These filters are designed to flag variants that are too close to an indel or each other, respectively, as these may stem from alignment artifacts and therefore be false positives. A value of 5 for these filters would mean that SNPs that are five or fewer bp from an indel call would be discarded. The same would apply to clusters of SNPs that are 5 or fewer bases apart from each other.

Review Question 3

Can you think of other possible filters that would need to be applied to the data post-variant calling to reduce the number of false positive calls?

These and other post-variant calling filters can be applied to a VCF file. Programs such as bcftools [5] or GATK VariantFiltration [4] can be used for this purpose, and their parameters can be tweaked to suit the researcher's needs. Typically, after these are run, the "FILTER" field in the VCF will be annotated as "PASS" if the variant has passed all specified filters, or will have specified the filters that it has failed. Usually, the subsequent analyses would be carried out only with those variants that passed all filters.

Now that we have seen the factors that researchers take into account when performing an NGS study, we will next discuss the types of analyses they can follow to identify *de novo* DNA variants or those that are likely to increase the risk of a disease.

10.5   *De novo* Genetic Variants: Population-Level Studies and Analyses Using Pedigree Information

De novo mutations are crucial to the evolution of species and play an important role in disease. *De novo* genetic variants are defined as those somatically arising during the formation of gametes (oocytes, sperm) or that occur postzygotically. Only the mutations present in germ cells can be transmitted to the next generation. Usually, when searching for *de novo* variants in children (usually affected by developmental disorders), researchers study trios, *i.e.* both parents and the child. The task is simplified because at 1.0×1.10^{-8} to 1.8×1.10^{-8} per nucleotide mutation rate, only a few *de novo* mutations are expected in the germline of the child (a range of 44–88 according to [20]). If more than one variant fulfills these criteria, bioinformatic methodologies such as examination of the extent of conservation throughout evolution, consequence prediction, and gene prioritization are then used to pinpoint the most likely gene variants underlying the phenotype. Functional studies such as cell growth experiments or luciferase assays can then be performed to demonstrate the biological consequences of the variant.

This type of filtering methodology has been extensively applied by projects such as the Deciphering Developmental Disorders (DDD) study [21]. This Consortium applied microarray and exome sequencing technologies to 1,133 trios (affected children and their parents) and was able to increase by 10% the number of children that could be diagnosed, as well as identifying 12 novel causative genes [22]. It may also be useful for the detection of causal genetic variation for neurodevelopmental disorders [23]. However, while tremendously useful in the cases where the three sequences are available, and where the variant is present in the child and not the parents, this strategy is not that useful in those cases where the causal variant may also be present in the parents or where it has a lower penetrance.

Another definition of a "*de novo*" variant may be one that has never before been seen in a population, which is identified through comparisons against population variation databases such as gnomAD [15] and dbSNP [24]. Sometimes, researchers assume that a rare variant, because it is rare (and perhaps because it falls in a biologically relevant gene) then it must underlie their phenotype of interest. However, this is nearly always not true: Depending on ancestry, estimates are that humans can carry up to 20,000 "singletons" (this is, genetic variants only observed once in a dataset) [25] and can carry more than 50 genetic variants that have been classified as disease-causing [15]. This point has been beautifully illustrated by Goldstein and colleagues [26]: They analyzed sequencing data from a control sample and reported finding genetic variants falling in highly conserved regions from protein-coding genes, that have a low allelic frequency in population databases, that have a strong predicted effect on protein function and in genes that can be connected to specific phenotypes in disease databases. However, even if fulfilling all these criteria, these variants clearly do not have a phenotype. They call the tendency of these kind of variants to be assumed as causal as "the narrative potential," which is unfortunately common in the literature [27]. Therefore, in the next section we will summarize the aspects that need to be taken into account in order to confidently assign a genetic variant as causative for a phenotype.

10.6 Filtering Genetic Variants to Identify Those Associated to Phenotypes

Given the huge number of genetic variants usually identified in NGS studies (12,000 in exomes, ~5 million in genomes) [28], filtering and post-processing to pinpoint candidates may be the most labor-intensive tasks out of the whole analysis pipeline. Depending on the researcher's biological question, they may need to tune these parameters to better answer it. For example, if they are searching for rare variation in pedigrees that may predispose to a disease, they may want to set quite permissive quality thresholds so as to not lose any potential candidates, but making sure that any potential variants are confirmed through resequencing by another orthogonal methodology such as capillary sequencing. If, on the other hand, they are analyzing a large cohort of individuals in order to describe patterns of variation, then they will need to be much stricter quality filters.

10.6.1 Variant Annotation

Variant annotation can help researchers filter and prioritize functionally important variants for further study. Several tools for functional annotation have been developed; some of them are based on public databases and are limited to known variants, while others have been developed for the annotation of novel SNPs.

Functional prediction of variants can be done through different approaches, from sequence-based analysis to structural impact on proteins. Predicted effects of identified variants can be assessed through tools such as Ensembl-VEP [29] and SnpEff [30]. On top of the predicted consequences on protein function (*e.g.*, whether a variant is missense, stop-gain, frameshift-inducing, etc.), these tools can also perform annotations at the level of allele frequency against public databases such as 1000 Genomes and GnomAD, whether the variant has been seen before either in populations or somatically in cancer (dbSNP and COSMIC annotations), whether it falls in an evolutionarily conserved site (GERP and PolyPhen-2 scores), and whether it has been found to have clinical relevance (ClinVar annotations), among others. Genomic region-based annotations can also be performed, referring to genomic elements other than genes, such as predicted transcription factor binding sites, predicted microRNA target sites, and predicted stable RNA secondary structures [31]. All these annotations can aid a researcher to focus on those variants predicted to be associated to their phenotype of interest.

However, these steps to identify variant candidates are only part of the story. As we mentioned above, even if the variants are real and seem to have an effect on gene function, this alone is not enough evidence to link the variant causally to a phenotype [32]. Researchers should be wary of any potential positive associations and should consider alternate hypotheses before reporting their identified variants as causal (or they may be publicly challenged, see, for example, [33, 34]).

10.6.2 Evaluating the Evidence Linking Variants Causally to Phenotypes

After these essential filtering and annotation steps have been performed, a researcher then needs to assess the amount of evidence supporting the potential causality of a genetic variant. The first line of evidence needs to be statistical: Assuming a candidate variant exists, the first question would be, how likely would it be to obtain an equivalent result by chance if any other gene were to be considered? For example, a 2007 study by Chiu and collaborators assumed that two novel missense genetic variants in the *CARD3* gene were causal of familial hypertrophic cardiomyopathy [35]. They assumed causality based on four criteria: If the variant had been seen in other cardiomyopathy patients, if it was absent from 200 alleles from controls, if it was conserved among species and isoforms and if it co-segregated with the disease in affected families. However, the chance of all these criteria being fulfilled by chance alone if any other genes had been considered is high—as a study subsequently found by assessing a larger gene panel and calculating the expected number of variants in the gene [34]. Additionally, both positive and negative evidence for the hypothesis should be carefully evaluated, for example, in the same cardiomyopathy study some of the "potentially causal" variants predicted by bioinformatics algorithms did not co-segregate with the phenotype [34]. The increasing availability of sequencing data in large cohorts such as gnomAD should help establishing causality as more accurate allele frequencies are reported per population [15]. This is an important point—allele frequencies should be matched by ancestry as closely as possible, as it is known that they can vary greatly among different populations [25].

Another important set of criteria, highlighted by MacArthur et al [32], argues that when analyzing potentially monogenic diseases, genes that have previously been confidently linked to similar phenotypes should be analyzed as the first potential candidates before proceeding to explore novel genes, and that if a researcher does proceed to analyzing further genes, then multiple independent carrier individuals must present with similar clinical phenotypes. Additionally, it is desirable that the distribution of variants in a suitable control population is examined, for example, if a researcher has identified a novel stop-gained variant in a candidate gene, how many other stop-gained variants are found in population-level variation catalogues?

Finally, statistical evidence and multiple computational approaches may strongly suggest that a variant is disease-causing. However, whenever possible, researchers should perform functional studies that indicate this is the case, whether by using tissue derived from patients themselves, cell lines, or model organisms. The comprehensive view provided by statistical, computational, and functional studies then may be enough for a researcher to report a potential causal variant. In doing so, it is recommended that all available evidence is detailed, clear and uncertain associations are reported and that all genetic data is released whenever possible [32].

10.6.3 Variant Filtering and Visualization Programs

Finally, visual representation of genomic data can be highly useful for the interpretation of results [28]. Visualization tools can help users browse mapped experimental data along with annotations, visualize structural variants, and compare sequences. These programs can be available as stand-alone tools or as web applications, and vary in the amount of bioinformatics knowledge necessary to operate them. Here we will review some of the most popular and that we consider useful, but there are many others suited for different purposes and with a range of functionalities.

- *Integrative Genomics Viewer (IGV)* [36] A very popular, highly interactive tool that is able to process large amounts of sequencing data in different formats and display read alignments, read- and variant-level annotations, and information from other databases. Website: http://software.broadinstitute.org/software/igv/
- *Galaxy* [37] Another highly popular, web-based platform that allows researchers to perform reproducible analyses through a graphical interphase. Users can load files in the FASTA, BAM, and VCF formats, among others, and perform data analysis and variant filtering in an intuitive way. Website: https://usegalaxy.org/
- *VCF/Plotein* [38] This web-based, interactive tool allows researchers to load files in the VCF format and interactively visualize and filter variants in protein-coding genes. It incorporates annotations from other external databases.
  Website: https://vcfplotein.liigh.unam.mx/

Take Home Message
- There are a number of different methods for performing variant calling, these can be naive, probabilistic, and heuristic. Probabilistic methods are the most widely used and implement a form of Bayes' Theorem. However, algorithm choice will depend on the researcher's study design.
- Sample storage and preparation methods may introduce errors that increase false positive calls and therefore should be considered when designing an analysis pipeline.
- Post-variant calling filters that analyze the distribution of variants across all sequencing reads will usually need to be applied to data in order to reduce false positive calls.
- True *de novo* genetic variants can be identified by analyzing trios with an affected child, in other scenarios a number of annotations and filtering steps need to be applied to identify candidate variants.
- For a researcher to ascribe phenotype causality to a genetic variant, the result of gene- and variant-level annotations are not enough; a number of further statistical, bioinformatic, and functional considerations need to be taken into account.
- Variant filtering and visualization tools can aid a researcher to perform the above mentioned steps in an easy and intuitive way.

Answers to Review Questions

Answer to Question 1: $Q = 34.77$.

Answer to Question 2: The logical option is for the researcher to go back and analyze, through tools such as *Samtools depth*, whether indeed there is enough coverage at every assessed site, and to mark it as "no call" otherwise. A novel VCF format, called gVCF and outputted by GATK, can now give reference call confidence scores.

Answer to Review Question 3: There are a number of filters already implemented in variant filtering tools, some of these are a threshold for Phred-scaled variant quality, minimum depth, whether the variant is in a low-complexity region (*i.e.,* a highly repetitive region that may increase alignment errors), and more.

Box: Genome Assembly

If there is no reference genome available for our species of interest, it may be worth trying to create one from scratch. To do this, DNA fragments of the targeted species are sequenced in high quantity, resulting in sequenced reads that theoretically cover the entire genome. Reads are aligned and merged based on their overlapping nucleotides, assembling long DNA sequences. When the order of bases is known to a high-confidence level, this genomic sequence will be named a "contig." Multiple contigs can be assembled together to form a scaffold based on paired read information. A scaffold is a portion of the genome sequences composed of contigs but which might contain gaps in between them.

There are various tools to close gaps between scaffolds. Scaffolds can then be joined together to form a chromosome. Despite how easy this may sound, genome assembly has its difficulties and it can vary between one organism to another (for example, an uneven representation of the genome due to sequencing sensitivity to GC bias, which can cause gaps between scaffolds) [39].

Further Reading

The GATK blog: https://software.broadinstitute.org/gatk/documentation/article.php?id=4148

- Goldstein DB, Allen A, Keebler J, Margulies EH, Petrou S, Petrovski S, et al. Sequencing studies in human genetics: design and interpretation. Nat Rev Genet. 2013 Jul;14 (7):460–70.
- Deciphering Developmental Disorders Study. Large-scale discovery of novel genetic causes of developmental disorders. Nature. 2015 Mar 12;519(7542):223–8.
- MacArthur DG, Manolio TA, Dimmock DP, Rehm HL, Shendure J, Abecasis GR, et al. Guidelines for investigating causality of sequence variants in human disease. Nature. 2014 Apr 24;508(7497):469–76.

Somatic variant callers

Mutect2. (Bayesian) Cibulskis, Kristian, Michael S. Lawrence, Scott L. Carter, Andrey Sivachenko, David Jaffe, Carrie Sougnez, Stacey Gabriel, Matthew Meyerson, Eric S. Lander, and Gad Getz. "Sensitive Detection of Somatic Point Mutations in Impure and Heterogeneous Cancer Samples." Nature Biotechnology. 2013;31(3):213–19.

GATK HaplotypeCaller (Bayesian). McKenna, Aaron, Matthew Hanna, Eric Banks, Andrey Sivachenko, Kristian Cibulskis, Andrew Kernytsky, Kiran Garimella, et al. "The

Genome Analysis Toolkit: A MapReduce Framework for Analyzing next-Generation DNA Sequencing Data." Genome Research. 2010;20(9):1297–1303.

Varscan2. (Heuristic). Koboldt, Daniel C., Qunyuan Zhang, David E. Larson, Dong Shen, Michael D. McLellan, Ling Lin, Christopher A. Miller, Elaine R. Mardis, Li Ding, and Richard K. Wilson. "VarScan 2: Somatic Mutation and Copy Number Alteration Discovery in Cancer by Exome Sequencing." Genome Research. 2012;22(3):568–76.

10.7   A Practical Example Workflow

A workflow chart for a typical variant calling analysis is shown in Fig. 10.3.

1. Preprocessing
   - Check that the base calling and read alignment are accurate using the standard Phred quality sore.
   - Know the expected ploidy in your experiment.
   - Know the type of variants you want to identify.
   - Know your sequencing platform.
   - Calculate the sequencing depth.
   - Check the coverage ratio between X and Y chromosome to determine sample sex concordance.
   - Identify duplicated or related samples.
   - Apply a filter on low-complexity regions according to your research interest.
   - Select the appropriate variant caller.

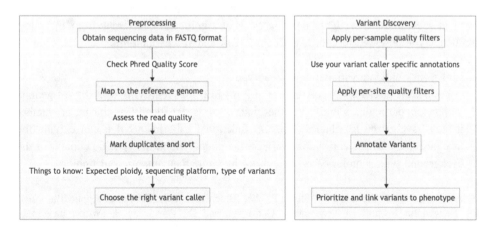

Fig. 10.3 Workflow-Chart for identification of genetic variants and de novo mutations

2. Once the VCF file is generated:
   - Apply per-sample filters that can relate to your cohort, some examples are:

```
#Check the missingness on a per-individual basis
vcftools -vcf *.vcf -missing-indv -out *
Check the heterozygosity rate of each sample
vcftools -vcf *.vcf -het -out *
# Check the p-value for each site from a Hardy-Weinberg Equilibrium test
vcftools -vcf *.vcf -hardy -out
```

After all these metrics are calculated, we suggest you graph each of them to easily identify outliers and define a threshold for further filtering. These metrics should also be calculated per variant site and filters should be applied under that dimension.

- Apply per-site filters that can relate to your variant calling method, for example, check the strand bias (identified by performing a Fisher test) "FS" and/or the strand OR "SOR" values.

3. To identify *de novo* variants
   Annotate your VCF file with the previously known information for each variant using tools like Ensembl-VEP [29] or SnpEff [30].

```
ensembl-vep/vep -i *.vcf -o *.vep -everything
SnpEff eff -v * -stats * refgenome *.vcf
```

Check for the allele frequency of your variants in the population that your samples came from in the different available data bases, is it significantly different from the allele frequency you observed in your experiment? How can you explain this?

4. Link your candidate variants to a phenotype
   Follow the advice by MacArthur et al 2014 [32] for identifying causality of genetic variants, in particular, identify whether your result is statistically significant or whether it may have arisen by chance. Perform functional experiments that can explain the mechanism by which your variant affects the phenotype in the specific context of the background your samples carry. Search for literature that support your findings.

Acknowledgements We thank Dr. Stefan Fischer (Biochemist at the Faculty of Applied Informatics, Deggendorf Institute of Technology, Germany), and Dr. Petr Danecek (Wellcome Sanger Institute, United Kingdom) for reviewing this chapter and suggesting extremely relevant enhancements to the original manuscript.

References

1. Ewing B, Green P. Base-calling of automated sequencer traces using Phred. II. Error probabilities. Genome Res. 1998;8(3):186–94.
2. Nielsen R, Paul JS, Albrechtsen A, Song YS. Genotype and SNP calling from next-generation sequencing data. Nat Rev Genet. 2011;12(6):443–51.
3. Xu C. A review of somatic single nucleotide variant calling algorithms for next-generation sequencing data. Comput Struct Biotechnol J. 2018;16:15–24.
4. McKenna A, Hanna M, Banks E, Sivachenko A, Cibulskis K, Kernytsky A, et al. The genome analysis toolkit: a map reduce framework for analyzing next-generation DNA sequencing data. Genome Res. 2010;20(9):1297–303.
5. Li H. A statistical framework for SNP calling, mutation discovery, association mapping and population genetical parameter estimation from sequencing data. Bioinformatics. 2011;27 (21):2987–93.
6. You N, Murillo G, Su X, Zeng X, Xu J, Ning K, et al. SNP calling using genotype model selection on high-throughput sequencing data. Bioinformatics. 2012;28(5):643–50.
7. Koboldt DC, Zhang Q, Larson DE, Shen D, McLellan MD, Lin L, et al. VarScan 2: somatic mutation and copy number alteration discovery in cancer by exome sequencing. Genome Res. 2012;22(3):568–76.
8. Cai L, Yuan W, Zhang Z, He L, Chou K-C. In-depth comparison of somatic point mutation callers based on different tumor next-generation sequencing depth data. Sci Rep. 2016;6:36540.
9. Prentice LM, Miller RR, Knaggs J, Mazloomian A, Aguirre Hernandez R, Franchini P, et al. Formalin fixation increases deamination mutation signature but should not lead to false positive mutations in clinical practice. PLoS One. 2018;13(4):e0196434.
10. Hayward NK, Wilmott JS, Waddell N, Johansson PA, Field MA, Nones K, et al. Whole-genome landscapes of major melanoma subtypes. Nature. 2017;545(7653):175–80.
11. Costello M, Pugh TJ, Fennell TJ, Stewart C, Lichtenstein L, Meldrim JC, et al. Discovery and characterization of artifactual mutations in deep coverage targeted capture sequencing data due to oxidative DNA damage during sample preparation. Nucleic Acids Res. 2013;41(6):e67.
12. Briggs AW, Stenzel U, Meyer M, Krause J, Kircher M, Pääbo S. Removal of deaminated cytosines and detection of in vivo methylation in ancient DNA. Nucleic Acids Res. 2010;38(6): e87.
13. Newman AM, Lovejoy AF, Klass DM, Kurtz DM, Chabon JJ, Scherer F, et al. Integrated digital error suppression for improved detection of circulating tumor DNA. Nat Biotechnol. 2016;34 (5):547–55.
14. Wang J, Raskin L, Samuels DC, Shyr Y, Guo Y. Genome measures used for quality control are dependent on gene function and ancestry. Bioinformatics. 2015;31(3):318–23.
15. Lek M, Karczewski KJ, Minikel EV, Samocha KE, Banks E, Fennell T, et al. Analysis of protein-coding genetic variation in 60,706 humans. Nature. 2016;536(7616):285–91.
16. Kosugi S, Momozawa Y, Liu X, Terao C, Kubo M, Kamatani Y. Comprehensive evaluation of structural variation detection algorithms for whole genome sequencing. Genome Biol. 2019;20 (1):117.
17. Bohannan ZS, Mitrofanova A. Calling variants in the clinic: informed variant calling decisions based on biological, clinical, and laboratory variables. Comput Struct Biotechnol J. 2019;17:561–9.
18. Danecek P, Auton A, Abecasis G, Albers CA, Banks E, DePristo MA, et al. The variant call format and VCFtools. Bioinformatics. 2011;27(15):2156–8.
19. Danecek P, Nellåker C, McIntyre RE, Buendia-Buendia JE, Bumpstead S, Ponting CP, et al. High levels of RNA-editing site conservation amongst 15 laboratory mouse strains. Genome Biol. 2012;13(4):26.

20. Acuna-Hidalgo R, Veltman JA, Hoischen A. New insights into the generation and role of de novo mutations in health and disease. Genome Biol. 2016;17(1):241.
21. Firth HV, Wright CF, Study DDD. The Deciphering Developmental Disorders (DDD) study. Dev Med Child Neurol. 2011;53(8):702–3.
22. Deciphering Developmental Disorders Study. Large-scale discovery of novel genetic causes of developmental disorders. Nature. 2015;519(7542):223–8.
23. Carneiro TN, Krepischi AC, Costa SS, Tojal da Silva I, Vianna-Morgante AM, Valieris R, et al. Utility of trio-based exome sequencing in the elucidation of the genetic basis of isolated syndromic intellectual disability: illustrative cases. Appl Clin Genet. 2018;11:93–8.
24. Sherry ST, Ward MH, Kholodov M, Baker J, Phan L, Smigielski EM, et al. dbSNP: the NCBI database of genetic variation. Nucleic Acids Res. 2001;29(1):308–11.
25. 1000 Genomes Project Consortium, Auton A, Brooks LD, Durbin RM, Garrison EP, Kang HM, et al. A global reference for human genetic variation. Nature. 2015;526(7571):68–74.
26. Goldstein DB, Allen A, Keebler J, Margulies EH, Petrou S, Petrovski S, et al. Sequencing studies in human genetics: design and interpretation. Nat Rev Genet. 2013;14(7):460–70.
27. Bell CJ, Dinwiddie DL, Miller NA, Hateley SL, Ganusova EE, Mudge J, et al. Carrier testing for severe childhood recessive diseases by next-generation sequencing. Sci Transl Med. 2011;3 (65):65ra4.
28. Pabinger S, Dander A, Fischer M, Snajder R, Sperk M, Efremova M, et al. A survey of tools for variant analysis of next-generation genome sequencing data. Brief Bioinform. 2014;15(2):256–78.
29. McLaren W, Gil L, Hunt SE, Riat HS, Ritchie GRS, Thormann A, et al. The Ensembl Variant Effect Predictor. Genome Biol. 2016;17(1):122.
30. Cingolani P, Platts A, Wang LL, Coon M, Nguyen T, Wang L, et al. A program for annotating and predicting the effects of single nucleotide polymorphisms, SnpEff: SNPs in the genome of Drosophila melanogaster strain w1118; iso-2; iso-3. Fly. 2012;6(2):80–92.
31. Wang K, Li M, Hakonarson H. ANNOVAR: functional annotation of genetic variants from high-throughput sequencing data. Nucleic Acids Res. 2010;38(16):e164.
32. MacArthur DG, Manolio TA, Dimmock DP, Rehm HL, Shendure J, Abecasis GR, et al. Guidelines for investigating causality of sequence variants in human disease. Nature. 2014;508 (7497):469–76.
33. Minikel EV, MacArthur DG. Publicly available data provide evidence against NR1H3 R415Q causing multiple sclerosis. Neuron. 2016;92(2):336–8.
34. Verhagen JMA, Veldman JH, van der Zwaag PA, von der Thüsen JH, Brosens E, Christiaans I, et al. Lack of evidence for a causal role of CALR3 in monogenic cardiomyopathy. Eur J Hum Genet. 2018;26(11):1603–10.
35. Chiu C, Tebo M, Ingles J, Yeates L, Arthur JW, Lind JM, et al. Genetic screening of calcium regulation genes in familial hypertrophic cardiomyopathy. J Mol Cell Cardiol. 2007;43(3):337–43.
36. Robinson JT, Thorvaldsdóttir H, Winckler W, Guttman M, Lander ES, Getz G, et al. Integrative genomics viewer. Nat Biotechnol. 2011;29(1):24–6.
37. Afgan E, Baker D, van den Beek M, Blankenberg D, Bouvier D, Čech M, et al. The Galaxy platform for accessible, reproducible and collaborative biomedical analyses: 2016 update. Nucleic Acids Res. 2016;44(W1):W3–10.
38. Ossio R, Garcia-Salinas OI, Anaya-Mancilla DS, Garcia-Sotelo JS, Aguilar LA, Adams DJ, et al. VCF/Plotein: visualization and prioritization of genomic variants from human exome sequencing projects. Bioinformatics. 2019;35(22):4803–5.
39. Pop M. Genome assembly reborn: recent computational challenges. Brief Bioinform. 2009;10 (4):354–66.

Design and Analysis of RNA Sequencing Data

11

Richa Bharti and Dominik G. Grimm

Contents

R. Bharti · D. G. Grimm (✉)
Technical University of Munich, Campus Straubing for Biotechnology and Sustainability,
Bioinformatics, Straubing, Germany

Weihenstephan-Triesdorf University of Applied Sciences, Straubing, Germany
e-mail: dominik.grimm@hswt.de

© Springer Nature Switzerland AG 2021
M. Kappelmann-Fenzl (ed.), *Next Generation Sequencing and Data Analysis*, Learning
Materials in Biosciences, https://doi.org/10.1007/978-3-030-62490-3_11

What You Will Learn in This Chapter
In this chapter, we introduce the concept of RNA-Seq analyses. First, we start to provide an overview of a typical RNA-Seq experiment that includes extraction of sample RNA, enrichment, and cDNA library preparation. Next, we review tools for quality control and data pre-processing followed by a standard workflow to perform RNA-Seq analyses. For this purpose, we discuss two common RNA-Seq strategies, that is a reference-based alignment and a *de novo* assembly approach. We learn how to do basic downstream analyses of RNA-Seq data, including quantification of expressed genes, differential gene expression (DE) between different groups as well as functional gene analysis. Eventually, we provide a best-practice example for a reference-based RNA-Seq analysis from beginning to end, including all necessary tools and steps on GitHub: https://github.com/grimmlab/BookChapter-RNA-Seq-Analyses.

11.1   Introduction

The central dogma of molecular biology integrates the flow of information encoded in DNA via transcription into RNA molecules that eventually translates into proteins inside a cell (Fig. 11.1, also see Chap. 1, Section 0). Any alteration, e.g. due to genetic, lifestyle, or environmental factors, might change the phenotype of an organism [1]. These alterations, e. g. copy number variations or mutational modifications of RNA molecules, affect the regulation of biological activities within individual cells [2, 3]. The entirety of all coding and non-coding RNAs derived from a cell at a certain point in time is referred to as the transcriptome [4]. Apparently, any change in the transcriptome culminates into functional alterations at both cellular and organismic level. Therefore, quantifying transcriptome variations and/or gene expression profiling remains crucial for understanding phenotypic alterations associated with disease and development [5, 6].

In the past, quantitative polymerase chain reaction (qPCR) was used as the tool of choice for quantifying transcripts and for performing gene expression analyses. Although qPCR remains a cheap and accurate technique for analyzing small sets of genes or groups of genes, it fails to scale to genome-wide level [7]. The introduction of DNA helped to scale transcriptomic studies to a genome-wide level due to their ability to accurately analyze thousands of transcripts at low cost [8, 9]. However, requirements of *a priori* knowledge of genome sequence, cross-hybridization errors, presence of artifacts, and the inability to analyze alternate splicing and non-coding RNAs limit the usage of microarrays [10, 11]. Currently, next-generation sequencing (NGS) has revolutionized the transcriptomic analysis landscape due to higher coverage, detection of low abundance and novel transcripts,

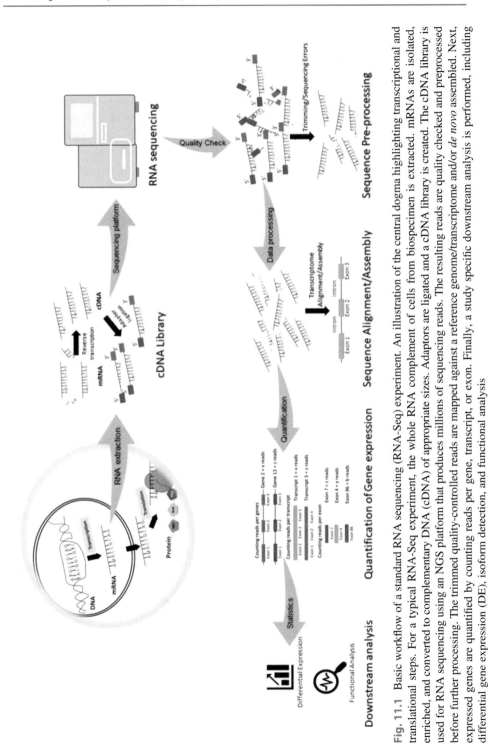

Fig. 11.1 Basic workflow of a standard RNA sequencing (RNA-Seq) experiment. An illustration of the central dogma highlighting transcriptional and translational steps. For a typical RNA-Seq experiment, the whole RNA complement of cells from biospecimen is extracted. mRNAs are isolated, enriched, and converted to complementary DNA (cDNA) of appropriate sizes. Adaptors are ligated and a cDNA library is created. The cDNA library is used for RNA sequencing using an NGS platform that produces millions of sequencing reads. The resulting reads are quality checked and preprocessed before further processing. The trimmed quality-controlled reads are mapped against a reference genome/transcriptome and/or *de novo* assembled. Next, expressed genes are quantified by counting reads per gene, transcript, or exon. Finally, a study specific downstream analysis is performed, including differential gene expression (DE), isoform detection, and functional analysis

dynamic changes in messenger RNA (mRNA) expression levels, analysis of genetic variants, splice variants, and protein isoforms [6]. Moreover, NGS based RNA sequencing, referred to RNA-Seq, could also be used for analyzing non-coding RNAs, microRNA (miRNA), small interfering RNA (siRNA), and other small RNA classes [11].

During the past decade the basic RNA-Seq protocol for analyzing different types and qualities of samples has continuously been modified and a variety of optimized protocols have been released [12, 13]. A basic workflow summarizing the major steps of a standard RNA-Seq analysis is shown in Figure 11.1. As already described in Chap. 3 the first step involves the isolation and purification of total RNA from a sample as well as the enrichment of target RNA. In this step, poly(A) capture is commonly utilized to selectively isolate polyadenylated mRNA molecules. Further, depletion of highly abundant ribosomal (rRNA) and transfer RNAs (tRNA) helps in mRNA enrichment (see Table 1.1). In the second step, chemical or enzymatic fragmentation of mRNA molecules into appropriate sizes (e.g., 300–500 bp for Illumina sequencing) followed by complementary DNA (cDNA) synthesis is performed. Next, adapter ligation to the 3' and 5' ends of the cDNA is done followed by the creation of a cDNA library. The third step is the actual sequencing step using a modern NGS technology that generates millions of sequencing reads. Reads are then quality checked, trimmed, and genome or transcriptome mapping is performed. Finally, a study specific downstream analysis of data is conducted to investigate and analyze differentially expressed genes and to perform an isoform identification or genome annotation [14]. In the following sections we will describe the main steps of a standard and best-practice RNA-Seq analysis.

11.2   RNA Quality

Extraction of high-quality RNA is a key step in any RNA-Seq analysis. Several RNA extraction and purification methods (RNA purification involves cell lysis, RNAse inhibition, DNA removal, and isolation of RNA) and commercial kits are available that show considerable variability in quality and yield of RNA [15, 16]. In general, variabilities in these steps together with other physical factors, such as the nature of the sample, its stability and organism determine the quality of the isolated RNA [17]. Therefore, the determination of RNA integrity numbers (RIN) via Agilent Bioanalyzer or TapeStation systems is a critical first step in obtaining meaningful gene expression data. The RIN score relies on the amount of 18S and 28S to assess in vitro RNA degradation and thus RNA quality. It is highly recommended to use only RNA for RNA-Seq library preparation with a RIN score between 6 and 10, the higher the better. In terms of lower RIN scores ask an experienced scientist what to do.

11.3   RNA-Seq Library Preparation

The basic steps in RNA-Seq library preparation include efficient ribosomal RNA (rRNA) removal from samples followed by cDNA synthesis for producing directional RNA-Seq libraries [18, 19]. Few of the widely used RNA-Seq library preparation kits include Illumina TruSeq (https://www.illumina.com/products/by-type/sequencing-kits/library-prep-kits/truseq-rna-v2.html), Bioo Scientific NEXTFlex (http://shop.biooscientific.com/nextflex-small-rna-seq-kit-v3/), and New England Biolabs NEB Next Ultra (https://www.neb.com/products/e7370-nebnext-ultra-dna-library-prep-kit-for-illumina#Product%20Information).
In general, after removal of rRNA fractions, the remaining RNA is fragmented and reverse transcribed using random primers with 5'-tagging sequences. Next, the 5'-tagged cDNA are re-tagged at their 3' ends by a terminal-tagging process to produce double-tagged, single-stranded cDNA. In addition, Illumina adaptor sequences are added using limited-cycle PCR that ultimately results in a directional, amplified library. Finally, the amplified RNA-Seq library is purified and is further utilized for cluster generation and sequencing.

11.4   Choice of Sequencing Platform

Several NGS platforms (see Chap. 4) have been successfully implemented for RNA-Seq analysis in the past few years. Currently the three most widely used NGS platforms for RNA-Seq are the Illumina HiSeq, Ion Torrent, and SOLiD systems [20]. Although the nucleotide detection methodology varies for each platform, they follow similar library preparation steps. Either sequencing platform generates between 10 and 100 million reads, with typical read lengths of 300–500 bp. However, more recent sequencing technologies, such as Pacific Biosciences (PacBio) and Oxford Nanopore Technologies (MinION) can sequence full-length transcripts on a transcriptome-wide scale and can produce long reads ranging between 700 and 2000 bp [21]. Moreover, single-cell RNA-Seq (scRNA-Seq) has emerged recently as a new tool for precisely performing transcriptomic analysis at the level of individual cells [22]. Different RNA-Seq experiments require different read lengths and sequencing depths. Importantly, data generated by different RNA-Seq platforms vary and might affect the results and interpretations. Thus, the choice of the sequencing platform is crucial and highly depends on the study design and the objectives. An overview of platform specific differences is summarized in Table 11.1 and Chap. 4.

11.5   Quality Check (QC) and Sequence Pre-processing

After the sequencing experiment, a quality check (QC) of raw reads is required to filter out poor-quality reads resulting from errors in library preparation, sequencing errors, PCR artifacts, untrimmed adapter sequences, and presence of contaminating sequences [24]. Presence of low-quality reads often affect the downstream processing and interpretation of obtained results. Several tools, including *FastQC* [25], *htSeqTools* [26], and *SAMStat* [27], to assess the quality

Table 11.1 Technical specifications of common sequencing platforms used for RNA-Seq experiments. Adapted from Quail A et al. 2012 [23]

Platform	Run time	Raw error rate (%)	Read length	Paired reads	Insert size	Sample required
Illumina MiSeq	27 h	0.8	≤150 b	Yes	≤700 b	50–1000 ng
Illumina GAIIx	10 days	0.76	≤150 b	Yes	≤700 b	50–1000 ng
Illumina HiSeq 2000	11 days	0.26	≤150 b	Yes	≤700 b	50–1000 ng
Ion Torrent PGM	2 h	1.71	~200 b	Yes	≤250 b	100–1000 ng
PacBio RS	2 h	12.86	~1500 b	No	≤10 kb	~1 µg

of raw sequencing reads. *FastQC* is a commonly used tool providing various analyses and quality assessments about the overall quality of the raw sequencing to identify low-quality reads (see Sect. 7.3.1). QC is followed by a pre-processing step that involves removal of low-quality bases, adapter sequences, and other contaminating sequences. *Cutadapt* [28] and *Trimmomatic* [29] are two popular tools utilized for filtering (removing adapter and/or other contaminating sequences) based on parameters that could be easily customized by the user.

Figure 11.2 summarizes all necessary steps schematically. In addition, we provide a detailed practical example for a reference-based RNA-Seq analysis on GitHub: https://github.com/grimmlab/BookChapter-RNA-Seq-Analyses. For this example, data from the Sequence Read Archive (SRA) [30] are used. Fastq files are downloaded and an initial quality assessment is done using *FastQC*:

```
fastqc \
SRR5858228_1.fastq.gz \
-o rawreads_QA_stats
```

```
Description of parameters:
SRR5858229_1.fasta.gz is the input example file for quality assessment
-o          is path to output directory, e.g reads_QA_stats
```

Next, reads with low-quality have to be filtered and adapters have to be removed. Optimal parameters for filtering are derived from the *FastQC* quality report (see Chap. 7, Sect. 7.3.1). The tool *Cutadapt* can be used as follows:

```
cutadapt \
-q 20 \
-minimum-length 35 \
-a "CTGTCTCTTATACACATCT" \
-o reads/filtered_reads/SRR5858228_1_trimmed.gz \
SRR5858229_1.fastq.gz \ #Input file name
> SRR5858229_1_cutadapt_stats.txt # Output stats for given file
```

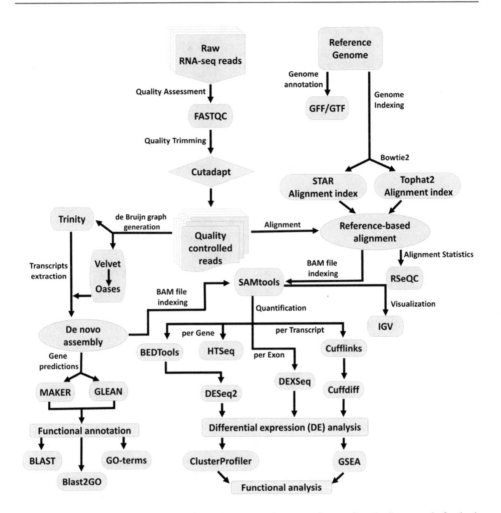

Fig. 11.2 A schematic illustration of a typical RNA-Seq experiment. Standard protocols for both reference-based alignment/mapping and *de novo* assembly are shown

```
Description of parameters:

-q                          phred score of 20

--minimum-length            the minimum sequence length (35)

-a                          the adapter sequence for trimming

-o                          the output file name
```

11.6 RNA-Seq Analysis

There are two main strategies for RNA-Seq analysis, one that is based on reference-based alignment and the second that is based on a *de novo* assembly. In addition, there are hybrid approaches that combine both strategies [4]. In the following, we will describe details for both, the reference-based alignment and the *de novo* assembly.

11.6.1 Reference-Based Alignment

A reference-based alignment is a strategy to align (map) individual reads (based on sequence similarities) against a chosen reference genome or a transcriptome sequence [31], as illustrated in Fig. 11.3. One aim of this approach is to quantitate transcript abundance at a genomic locus [32]. Here, each identified mapping location must indicate the origin of transcript as well as the total number of reads corresponding to the same transcript present in the dataset.

Reference-Genome Based Alignment A reference genome sequence is utilized for aligning the reads. For an accurate mapping it is important to consider that reads might originate from either cDNA corresponding to spliced transcripts or from non-spliced transcripts [33, 34]. In the case of spliced transcripts, contiguous read sequences are separated by intervening splice junction (SJ) boundaries and are split into two fragments and assigned separately. Thus, there are two main alignment approaches for aligning reads against a reference genome, that is spliced alignment and unspliced alignment [35, 36]. In addition, the alignment of reads against a reference genome leads to unaligned regions and gaps. Apparently, splice-aware aligners are generally suitable for most reference-genome based alignments. As the information on SJs in the reference genome is crucial, most aligners also refer to known annotated SJ databases to confirm the presence or absence of splice sites [37]. In addition, most aligners also perform sequence matching of terminal nucleotides in aligned reads with donor–acceptor sites of known splicing sites. Currently a number of splice-aware alignment tools are available, such as *BBMap* [38], *STAR* [39], *GMAP* [40], and *TopHat2* [41]. *STAR* utilizes a sequential maximum mappable seed search algorithm that consolidates into seed clustering and stitching steps to generate alignments. In contrast, *TopHat2* incorporates an exhaustive initial step where alignments are scanned for possible exon–exon junctions, which are eventually utilized in the final step for generating the final alignment. Importantly, parameters of alignment tools have to be carefully chosen to gain optimal alignment results.

Reference-Transcriptome Based Alignment This alignment method is mainly utilized when a well-annotated transcriptome or a set of known transcripts are available. Here, sequencing reads can be directly aligned in a continuous, error-free manner without involving computationally intensive steps, due to the availability of splicing information for reference transcripts [42]. For this purpose, aligners that are not splice-aware or

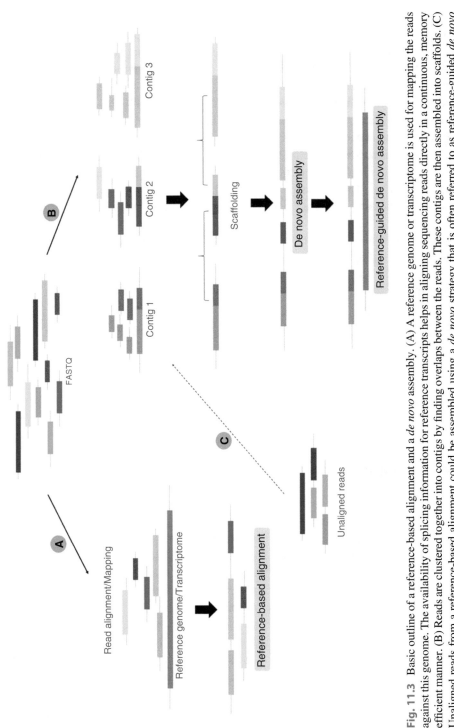
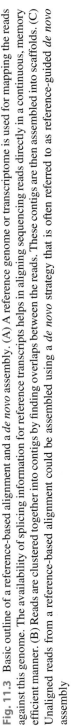

Fig. 11.3 Basic outline of a reference-based alignment and a *de novo* assembly. (A) A reference genome or transcriptome is used for mapping the reads against this genome. The availability of splicing information for reference transcripts helps in aligning sequencing reads directly in a continuous, memory efficient manner. (B) Reads are clustered together into contigs by finding overlaps between the reads. These contigs are then assembled into scaffolds. (C) Unaligned reads from a reference-based alignment could be assembled using a *de novo* strategy that is often referred to as reference-guided *de novo* assembly

ungapped aligners are used, since the alignment process does not essentially incorporate a range of gaps. However, as the alignment process utilizes a relatively reduced reference size in the form of selected transcriptomic sequences, it is not particularly useful for identifying novel exons, their expression patterns or splicing events [43, 44]. This is simply because all the available isoforms are aligned to the same exon of a gene multiple times, due to pre-existing splicing information. *Bowtie2* [45] and *MAQ* [46] are the two most widely used aligners for transcriptome-based alignment/mapping. *Bowtie2* utilizes a Burrows–Wheeler Transform (BWT) based FM index (Full-text index in Minute space) to perform fast and memory efficient alignments. On the other hand, *MAQ* identifies ungapped matches with low mismatch scores and assign them to a threshold-based mapping algorithm. This generates phred-scaled quality scores for each read alignment and a final alignment is produced by collating the best scoring matches for every read.

Reference genomes, e.g. for mouse, together with annotations can be downloaded from GENCODE [47] using the following command:

```
SOURCE_MOUSE=ftp://ftp.ebi.ac.uk/pub/databases/gencode/Gencode_mouse/
release_M24/
wget -c \
${SOURCE_MOUSE}/GRCm38.p6.genome.fa.gz
wget -c \
${SOURCE_MOUSE}/gencode.vM24.annotation.gff3.gz
wget -c \
${SOURCE_MOUSE}/gencode.vM24.annotation.gtf.gz
```

11.6.1.1 Choice of Reference-Based Alignment Program

The choice of a reference-based aligner for RNA-Seq data depends mainly on the splicing information or simply the type of organism [48]. For RNA-Seq reads originating from organisms without introns, that is prokaryotes/archaea, ungapped aligners or aligners that are not splice-aware are utilized. However, if the reads are aligned to intronic genomes (eukaryotic), splice-aware aligners like *TopHat2* [41], *STAR* [39], or *BBMap* [38] are used that incorporate a three-step alignment procedure. In the first step, sequencing reads are aligned to a reference genome or transcriptome sequence. Next, overlapping reads present at each locus are collated into a graph representing all isoforms. In the final step, the graph is resolved, and all isoforms associated with individual genes are identified [35]. In our example (see GitHub and Fig. 11.2), we use the commonly used tools *TopHat2* and *STAR*.

TopHat2: As mentioned in Chap. 9, Sect. 9.2.3.4 *TopHat2* is a fast and memory efficient tool that utilizes *Bowtie2* for alignment. Genomic annotations are used together with the available transcriptome information to produce precise spliced alignments. In addition, *TopHat2* allows excluding pseudogene sequences and shows minimum tolerance to mismatches and non-alignment of low-quality bases. *TopHat2* incorporates a multistep alignment algorithm that consists of an optional transcriptional alignment followed by a genomic alignment and finally a spliced alignment based on the available annotations (Fig. 11.4a). The main steps using *TopHat2* are summarized below:

Building Reference Index First, the reference genome of choice is indexed using *Bowtie2*. The generated index file and genomic FASTA file are then used for the genomic alignment of the reads.

```
bowtie2-build \
-f data/mouse-gencode-version-24/GRCm38.p6.genome.fa \
/data/TOPHAT_Genome_Index/ \
-p 8
```

```
Description of parameters:
-f      the path to genome Fasta file
-p      to launch a specified number of parallel threads
```

Read Alignment/Mapping Quality checked reads in *FASTQ* format or plain FASTA files are accepted as input for *TopHat2*. *TopHat2* is able to handle either single-end or paired-end reads and is able to integrate single-end reads into a paired-end alignment. Separate commands are used for utilizing genome index and transcriptome index files. Note, for an optimal pair-end alignment it is crucial to keep the same order of reads in the two files.

Availability of genomic annotations in the GTF/GFF file format could be utilized for initial transcriptome alignment. Transcriptome alignment involves the creation of a transcriptome index utilizing the genome index and the annotation information from GTF/GFF file.

The alignment using *TopHat2* is performed as follows:

```
tophat \
-o /analysis/mapping/tophat \
-p 8 \
-G data/mouse-gencode-version-24/gencode.vM24.annotation.gtf \
/data/TOPHAT_Genome_Index/GRCm38.p6.genome \
reads/filtered_reads/SRR5858228_1_trimmed.gz
```

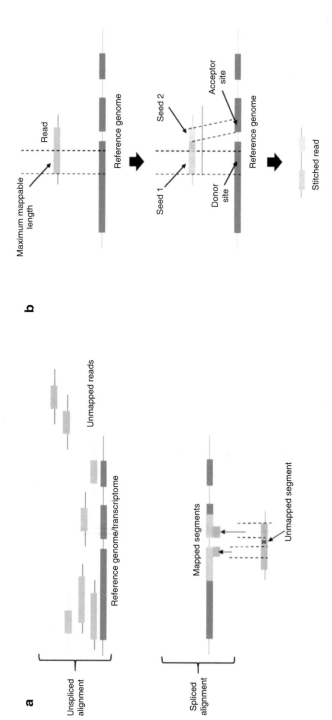

Fig. 11.4 Schematic demonstrating *TopHat2* and *STAR* alignment algorithms. (**a**) *TopHat2*: First an unspliced alignment of reads with exon sequences is performed, followed by a spliced alignment. In spliced alignment, reads that did not align to the reference genome/transcriptome are split into shorter segments and aligned again. A gapped alignment is performed in order to identify potential splice sites in the genomic region and matching segments are stitched together. (**b**) *STAR* utilizes maximum mappable length of a read that aligns. Here, the alignment involves splitting individual reads into pieces or seeds and identifying the best segment that can be mapped for each seed. Finally, the seeds are stitched together and mapped using genome sequence as uncompressed suffix arrays containing information on splice sites

```
Description of parameters:
-o          output directory
-p          the number of threads to align reads (default used is 8)
-G          the path to the annotation file
/data/TOPHAT_Genome_Index/GRCm38.p6.genome is the path and prefix name of reference genome file
reads/filtered_reads/SRR5858228_1_trimmed.gz is the path to preprocessed and trimmed reads
```

Output The alignments are stored in BAM file format. Additionally, a text file is generated indicating the alignment rate and the number of reads or pairs with multiple alignments. Further, identified exon junctions, insertions, and deletions are stored in a separate BED file.

   STAR: *STAR* (Spliced Transcripts Alignment to a Reference) is a fast alignment algorithm albeit having relatively higher memory requirements [39]. It is based on maximum mappable length approach that involves splitting individual reads into pieces or seeds and identifying best segments that can be mapped for each seed (Fig. 11.4b). Next, segments are stitched together and mapped using the genome sequence as an uncompressed suffix array including information of splice sites (see Chap. 9, Sect. 9.2.3.1). The main steps in alignment/mapping using *STAR* are described below:

Building/Exporting Reference Index *STAR* provides two options to either create a custom genome index using suffix arrays or directly utilize available *STAR* references indices. To include SJ annotation into the mapping process, a separate splice junction reference file is required while constructing a customized reference index. Additionally, the memory requirement for running *STAR* can be reduced by using a relatively sparser suffix array that eventually reduces alignment speed. The reference index is created as follows:

```
STAR \
-runThreadN 8 \
-runMode genomeGenerate \
-genomeDir data/STAR_Genome_Index \
-genomeFastaFiles data/mouse-gencode-version-24/GRCm38.p6.genome.fa \
-sjdbGTFfile data/mouse-gencode-version-24/gencode.vM24.annotation.gtf
```

```
Description of parameters:
--runThreadN           number of threads
--runMode              is set to genomeGenerate, which generates genome files
--genomeDir            the path to the output directory name
--genomeFastaFiles     the path to the genome fasta file
--sjdbGTFfile          the path to the annotation file
```

Read Alignment/Mapping The mapping process for *STAR* may either use an already available SJ annotated genome index or custom index built by the end-user. Following this, the FASTQ files are specified and the alignment is done based on default or user-defined parameters. Several user-defined parameters, such as inserting SAM attributes, identifying mismatches, SJ information, and file compression are available in *STAR*. The alignment can be executed as follows:

```
STAR \
-runMode alignReads \
-runThreadN 8 \
-genomeDir data/STAR_Genome_Index \
-readFilesIn reads/filtered_reads/SRR5858228_1_trimmed.gz \
-readFilesCommand zcat \
-sjdbGTFfile data/mouse-gencode-version-24/gencode.vM24.annotation.gtf \
-outFileNamePrefix /analysis/mapping/star/SRR5858228_1_trimmed \
-outSAMtype BAM SortedByCoordinate
```

```
Description of parameters:
--runMode              alignReads is set to map the reads against the reference genome
--runThreadN           the number of threads (default is 8)
--genomeDir            the path to the genome index directory
--readFilesIn          the path to trimmed reads
--readFilesCommand     for gzipped files (*.gz) use zcat
--sjdbGTFfile          the path to the annotation file
--outFileNamePrefix    output prefix name with its path
--outSAMtype           output sorted by coordinate
```

Output The alignments are stored in the SAM file format along with several other output files that include information on alignments, SJs, log files indicating run progress, read pairs, and mapping statistics.

Review Question 1

What is the difference between a splice aware and splice unaware alignment?

11.6.2 *De novo* or Reference-Free Assembly

De novo or reference-free assembly is performed if a reference genome is not available [49]. This involves creation of contigs mainly using an overlap-based collation of sequencing reads along with several other parameters (Fehler! Verweisquelle konnte nicht gefunden werden.). For gene expression analysis, the sequencing reads are mapped on the assembled transcriptome further leading to annotation and functional analysis [50]. The biggest advantage of *de novo* assembly is its non-dependence on a reference genome. It is advantageous in analyzing data for sequences originating from un-sequenced genomes, partially annotated, unfinished genome drafts, and new/rare genomes including that of epidemic/pandemic viruses such as Ebola, Nipah, and Covid-19. In most cases, the RNA-Seq analysis along with a *de novo* assembly provides first-hand information on transcripts and phylogeny. On the contrary, performing a *de novo* assembly in conjugation with a reference-based alignment could also help detecting novel transcripts and isoforms [51, 52]. Importantly, identification of novel splicing sites or transcripts via a *de novo* assembly does not essentially require alignment information of pre-existing splice sites. Another important advantage of a *de novo* assembly is its compatibility to both, short read and long read sequencing platforms in contrast to a reference-based alignment that prefers the former.

There are several popular *de novo* assembly tools available that include *Rnnotator* [53], *Trans-ABySS* [54], and *Trinity* [55]. Besides, alternative and efficient approaches are hybrid methods which use both, reference-based alignments and *de novo* assemblies [52] ((PMID: 29643938) Fehler! Verweisquelle konnte nicht gefunden werden., **right**).

11.6.2.1 Choice of de novo Assembly Tools

Currently, *de Bruijn* graphs are the most widely used algorithm for performing a *de novo* assembly [56]. Hence, most of the popular *de novo* or reference-free assembly tools for RNA-Seq data utilized *de Bruijn* graphs, including *Velvet*, *Oases*, and *Trinity*. *Velvet* and *Oases* are used simultaneously [57, 58]. *Velvet* is a genome assembly tool that generates assembly graphs that are further analyzed by *Oases* for finding paths in the graphs to identify transcript isoforms and to generate a draft assembly. *Trinity* is based on three main modules that perform an initial assembly and clustering, create individual *de Bruijn* graphs for each cluster, and finally extract sequences representing transcript isoforms present at enlisted gene locus [55]. Next, we describe how to use both *Velvet/Oases* and *Trinity* [49].

***Velvet* and *Oases*:** *Velvet* is a genome assembler that calculates *k-mers* of data and assigns the contigs into a *de Bruijn* graph. Similarly, *Oases* performs transcript assemblies utilizing the output of *Velvet*. It segments the graphs generated by *Velvet* into transcript isoforms linked to each locus. Both tools process single-end reads as default; however, *Velvet* supports paired-end reads using a single file containing adjacently located read pair. The main steps in a *de novo* assembly using *Velvet* and *Oases* are described below:

Creating de Bruijn graph The input data format is defined (FASTA/FASTQ; single end/ paired end) and the data is clustered based on a particular *k-mer* length (e.g., 20). A hash table is created which is utilized to generate a *de Bruijn* graph for the defined *k-mer* size. In this step, a hash table is created with defined k-mer size and then graph traversal is done to create *de Bruijn* graphs by *Velvet* as follows:

```
velveth vdir 25 -fastq -shortPaired chr18_12.fq
velvetg vdir -ins_length 200 -read_trkg yes
```

```
Description of parameters:
vdir              the name of output directory
25                k-mer size
-shortPaired      paired end reads
-fastq            reads in FASTQ format
Chr18_12.fq       input file name
-ins              defined insert size (fragment length)
-read_trkg        read tracking information for Oases
```

Contig Generation In the second step, the graph traversal and contig extraction are performed on the resulting *de Bruijn* graph. Here, the minimum transcript length and insert size (for paired-end reads) are pre-defined. Importantly, several different assemblies with varying *k-mer* lengths can be also performed simultaneously to obtain an optimal assembly. In this step, the resulting *de Bruijn* graphs are extracted into contigs for a defined transcript length using *Oases*:

```
oases vdir -ins_length 200 -min_trans_lgth 200
```

```
Description of parameters:
vdir              the name of input directory containing velvet output
-ins_length       defined insert size for paired end reads
-min_trans_lgth   minimum transcript length
```

Output The output is generated as a FASTA file containing all identified transcript sequences with locus, isoform information, confidence value between 0 and 1, and transcript length. For each *k-mer* defined previously, a separate result is generated that contains the corresponding assembly. In case of multiple assemblies with different *k* values, *Velvet* can process each assembly individually.

   Trinity: *Trinity* is a combination of three independent software modules, that is *Inchworm*, *Chrysalis*, and *Butterfly* that are applied sequentially to perform a *de novo* assembly of transcriptomes from RNA-Seq data [55]. *Trinity* initially partitions the RNA-Seq reads into a number of individual *de Bruijn* graphs, each representing transcriptional complexity at particular gene/locus. Next, each of these *de Bruijn* graphs are separately extracted into full-length splicing isoform for cataloguing different transcripts obtained from paralogous genes. The main steps in a *de novo* assembly using *Trinity* are described below:

Contig Generation First *Inchworm* extracts all overlapping *k-mers* from the RNA-Seq reads. Second, each unique *k-mer* is examined in decreasing order of abundance and transcript contigs are generated using a greedy extension algorithm based on *(k-1)-mer* overlaps. Finally, unique portions of alternatively spliced transcripts are saved for the next step. *Trinity* is executed from the command line using a single Perl script "Trinity.pl":

```
# For single-end reads
$TRINITY_HOME/Trinity.pl \
-seqType fq \
-single single.fq \
-max_memory 20G
# For paired-end reads
$TRINITY_HOME/Trinity.pl
-seqType fq \
-left left.fq \
-right right.fq \
-max_memory 20G
```

```
Description of parameters:
--seqType        the input sequence in FASTQ format (can be fa, or fq)
--max_memory     the suggested max memory to be use by Trinity in Gb of RAM
# if single reads:
    --single     single reads, one or more file names, comma-delimited
# if paired reads:
    --leftleft reads, one or more file names (separated by commas)
    --right      right reads, one or more file names (separated by commas)
```

Creating de Bruijn Graph In the next step, the generated contigs are clustered using *Chrysalis* based on regions originating from alternatively spliced transcripts or closely related gene families. Following this, a *de Bruijn* graph for each cluster is created and the reads are partitions among these contig clusters. These contig clusters are termed as "components" and the partitioning of RNA-Seq reads into 'components' helps to process large sets of reads in parallel.

Output In the final step, *Butterfly* processes individual *de Bruijn* graphs in parallel by tracing RNA-Seq reads through each graph and determining connectivity based on the read sequence. This results in reconstructed transcript sequences for alternatively spliced isoforms along with transcripts that correspond to paralogous genes. The final output is a single FASTA file containing reconstructed transcript sequences.

11.7 Functional Annotation of *de novo* Transcripts

Functional annotation of the *de novo* transcripts involves identifying biological information, such as metabolic activity, cellular and physiological functions of predicted genes, or gene products/proteins [59]. In general, functional annotation can be either performed using conventional homology search or using a gene ontology (GO-term) based mapping [60, 61]. In the homology search, closely related protein sequences are initially identified by using a *BLASTp UniProtKB* database search [62] and based on protein domains using the *Pfam* database [63]. After integrating *BLASTp* and *Pfam* outputs, remaining functional annotation is done using *BLASTx* and *HMMER* (http://hmmer.org/). Additionally, *Rnammer* [64] and *SignalP* [65] are used for predicting ribosomal RNA and signal peptide sequences, respectively.

In the gene ontology-based annotation, GO-terms associated with hits obtained from *Blast* results are retrieved and catalogued into biological process ontologies, molecular function ontologies, or cellular component ontologies. The ontology data provides information about the functions and physiological activities of identified gene products. Another widely used tool, *Blast2GO* [66] utilizes statistics of GO-term frequencies for analyzing the enrichment of GO annotations. Additionally, the Kyoto Encyclopedia of Genes and Genomes (KEGG) [67] pathways are used for predicting interactions between gene products and related metabolic activities [32].

11.8 Post-alignment/assembly Assessment and Statistics

After alignment and read mapping, an assessment is done to analyze the quality of the alignment based on information, such as total number of processed reads, % of mapped reads, or SJs identified with uniquely mapped reads. Before any downstream analysis can be done, the output files require some post-processing, including file format conversions (SAM/BAM), sorting, indexing, and merging. There are a variety of tools available for post-processing, that is *SAMtools* [68], *BAMtools* [69], *Sambamba* [70], and *Biobambam* [71]. *SAMtools* mainly incorporates methods for file conversions from SAM- into BAM-format or vice versa. This is important because the BAM file format is one of the main file formats for several downstream tools. Besides, *SAMtools* is frequently used for sorting and listing alignments in BAM files, e.g. based on mapping quality and statistics. Many of these tasks are summarized in our RNA-Seq workflow available on GitHub: https://github.com/grimmlab/BookChapter-RNA-Seq-Analyses.

The reference-based alignment statistics can be computed as follows:

```
python3 bam_stat.py \
-q 30 \
-i analysis/mapping/star/SRR5858228_1_trimmed.bam > \
analysis/mapping/star/SRR5858228_1_trimmed_RSeQC_alignment_stats.txt
```

```
Description of parameters:
-i          input bam file with its path
-q          mapping quality to determine uniquely mapped read
```

Finally, *RSeQC* generates a table in the respective **mapping** folder where unique reads are considered if their mapping quality is more than 30 [72].

11.9   Visualization of Mapped Reads

The RNA-Seq analysis of diverse datasets is usually automated. However, it still requires additional and careful interpretations depending on the study design. Apparently, visualization of read alignments helps to gain novel insights into the structure of the identified transcripts, exon coverages, abundances, identification of indels and SNPs as well as of splicing junctions (SJs) [55]. In fact, visualizing aligned or assembled reads in a genomic or transcriptomic context helps comparing and interpreting the obtained data together with reference annotations. Currently, several genome browsers provide visualization of sequencing data, including *JBrowse* [73], Integrative Genomics Viewer (*IGV*) [74], *UCSC* [75], and the *Chipster* [76] genome browser. In our RNA-Seq workflow example we use the *IGV* browser, as shown in (Fig. 11.2). *IGV* can handle and visualize genomic as well as transcriptomic data. It incorporates a data-tiling approach to support large datasets and a variety of file formats. In this approach, any user-entered genome data is divided into tiles that represent individual genomic regions.

11.10   Quantification of Gene Expression

Both reference-based alignments and *de novo* assemblies provide comprehensive information about read location and abundance that can be used for quantifying gene expression [77]. For a typical RNA-Seq experiment, an estimation of the total number of mapped reads can directly provide information on the number of transcripts. However, it may not be always true as eukaryotic gene expression involves alternative splicing, which generates several isoforms from the same gene. Importantly, these isoforms can have overlaps in the exon sequences and thus affect precise mapping and quantification. Hence, depending on the experimental design, gene expression quantification can be done by either counting reads per genes, transcripts, or exons to obtain error-free estimations [78].

In our RNA-Seq workflow, quantifications are generated from a reference-based RNA-Seq alignment using the following *shell* script (see GitHub repository):

```
./quantification.sh
```

Following this, one of the three quantification methods can be chosen as follows:

```
QUANT_COUNT=per_gene_bedtools #per_gene_htseq
or per_transcript_cufflinks
or per_exon_dexseq
```

11.11 Counting Reads Per Genes

The simplest method to quantify gene expression is to count the number of reads that align to each gene in the RNA-Seq data. Various tools such as *HTSeq* [79] and *BEDTools* [80] are available that can count reads per gene. The workflow on GitHub contains reads per gene counts for both *HTSeq* and *BEDTools*. Note, we will only use *HTSeq* for the following explanations.

HTSeq—is a Python-based tool that has several pre-compiled scripts for analyzing NGS data. The *HTSeq*-count script takes as input the genomic read alignments in SAM/BAM format together with genome annotations in GFF/GTF format. The algorithm matches exon locations listed in GFF/GTF file and counts reads mapped to that location. The output is generated by clustering the total number of exons for each gene along with information on number of reads that were left-out. The criterion for left-out or unmapped reads include alignments at multiple locations, low alignment qualities, ambiguous alignments, and no alignments or overlaps. This step is implemented in the described RNA-Seq analysis as follows:

```
htseq-count \
-f bam \
-a 10 \
-m intersect-strict \
-s no \
-t exon \
-i gene_id \
analysis/mapping/star/SRR5858228_1_trimmed.bam \
data/mouse-gencode-version-24/gencode.vM24.annotation.gtf > \
analysis/quantification/htseq-count/SRR5858228_1_trimmed_counts.csv
```

```
Description of parameters:
-f      format of the input data.
-a      will skip all reads with MAPQ alignment quality lower than the given
minimum value (default: 10).
-m    mode to handle reads overlapping more than one feature. In this case its
intersect-strict.
-s      is set whether the data is from a strand-specific assay. For
stranded=no, a read is considered overlapping with a feature regardless of
whether it is mapped to the same or the opposite strand as the feature
-t   feature type (3rd column in GFF file) to be used, all features of other
types are ignored (default, suitable for RNA-Seq analysis using an GENCODE GTF
file: exon)
-I      attribute to be used as feature ID. The default, suitable for RNA-Seq
analysis using an GENCODE GTF file, is gene_id.
```

Output All the outputs are stored in the folder: **analysis/quantification/htseq-count**. Finally, all *HTSeq* output files are combined to created final file **Read_per_features_combined.csv** in the same folder which will be used as input for differential expression analysis.

11.12 Counting Reads Per Transcripts

Quantification of gene expression, by counting reads per transcript, utilizes algorithms that precisely estimate transcript abundances. Subsequently, sequence overlaps present in multiple transcript isoforms makes gene assignment tricky in this case. Thus, an expectation maximization (EM) approach is generally utilized where initially reads are assigned to different transcripts according to their abundance followed by sequential modification of the abundances based on assignment probabilities. Several programs are available for estimating transcript abundances in RNA-Seq data, including *Cufflinks* [81] and *eXpress* [82]. In our RNA-Seq workflow (on GitHub), we used *Cufflinks,* which is explained in more detail in the section below.

Cufflinks utilizes a batch EM algorithm, where genomic alignments are accepted in BAM format and annotations in GTF file. A likelihood function accounts for sequence-specific biases and utilizes the abundance information to identify unique transcripts. A continuous re-estimation of abundances is done based on omission of sequence biases in each cycle. The count data output contains information on transcripts and genes as FPKM-tracking files with FPKM (Fragments Per Kilobase Million) values and associated confidence intervals. This is implemented in the described RNA-Seq workflow as follows:

```
cufflinks \
-G data/mouse-gencode-version-24/gencode.vM24.annotation.gtf \
-b data/mouse-gencode-version-24/GRCm38.p6.genome.fa \
-p 8 \
analysis/mapping/star/SRR5858228_1_trimmed.bam \
-o analysis/quantification/cufflinks-count
```

```
Description of parameters:
-G          path to the annotation file
-b          path to the genome fasta file
-p          number threads
-o          name of the output directory
```

11.13 Counting Reads Per Exons

In general, a single exon can appear multiple times in a GTF file, due to exon sequence overlaps associated with transcript isoforms. Thus, abundance estimation for exons involves cataloguing a set of nonoverlapping exonic regions. For quantifying gene expression using read counts per exon, the Bioconductor package *DEXSeq* [83] is mainly used.

DEXSeq modifies the input GTF file into a list of exon counting bins that list single exons or a part of an exon that overlaps. Alignment is performed in SAM format with data sorted by read names and/or chromosomal coordinates and a modified GTF file with exon counting bins is generated. The output count file contains the number of reads for every exon counting bin. Here, a list of non-counted reads is generated based on the criterion that includes unaligned reads, low-quality alignments, ambiguous or multiple overlaps.

DEXSeq is implemented in the described RNA-Seq analysis workflow in two steps. The first step is the preparation of annotations:

```
python2 dexseq_prepare_annotation.py \
data/mouse-gencode-version-24/gencode.vM24.annotation.gtf \
analysis/quantification/dexseq-count/DEXSEQ_GTF_annotation.gff
```

The second step performs the read counting:

```
python2 dexseq_count.py \
-p no \
-s no \
-r name \
analysis/quantification/dexseq-count/DEXSEQ_GTF_annotation.gff \
-f bam \
-a 10 \
analysis/mapping/star/SRR5858228_1_trimmed.bam \
analysis/quantification/dexseq-count/SRR5858228_1_trimmed_exon_counts.
csv
```

```
Description of parameters:
-p      for paired-end sequencing data, enter option -p yes, otherwise no
-s      if the library preparation step doesn't preserve strand information,
enter option -s no
-r      indicates whether the data is sorted by alignment position or by read
name
-f      input read format
-a      specifies minimum alignment quality. All reads with a lower quality
than specified (default -a 10) are skipped.
```

Review Question 2

Are there any differences between gene expression quantification methods?

11.14 Normalization and Differential Expression (DE) Analysis

A majority of RNA-Seq experiments are performed to obtain information about transcriptional differences among a set of samples (organisms, tissues, or cells) and conditions or treatments [84, 85]. Thus, to prevent errors in estimation of expression or transcriptional differences, normalization remains a critical step for a given RNA-Seq analysis. Normalization helps in rectifying errors in factors that affect preciseness of read mapping including read length, GC-content, and sequencing depth [86]. However, errors in normalization might generate large number of false positives that can eventually affect preciseness of these downstream analyses [87]. In general, RNA-Seq data normalization involves transformation of the read count matrix for obtaining correct comparisons of read counts across samples. Correct normalization generates correct relationships between normalized read counts, thus affecting analysis across different conditions/treatment across samples [88]. Although it was not deemed a necessary factor initially, modern RNA-Seq analysis, including differential expression (DE) analysis, highly depends on data normalization

[89]. Normalization protocols together with statistical testing might have the largest impact on the results of an RNA-Seq analysis [90, 91]. In this context, several normalization methods have been developed based on different assumptions in RNA-Seq experiments and corresponding gene expression analysis. There are mainly three major normalization strategies, that is normalization by library size, normalization by distribution, and normalization by controls [92]. In case of normalization by library size, differences in sequencing depth are removed by simply dividing by the total number of reads generated for each sample. Similarly, normalization by distribution involves equilibrating expression levels for non-DE genes, if the technical effects or treatments remain the same for DE and non-DE genes. Importantly, assumptions made by any normalization method should be always considered before choosing a preferable method that suits the study design [93]. For instance, normalization by library size should be chosen when the total mRNA remains equal or in the same range across different treatments/conditions despite any asymmetry. In contrast, normalization by distribution is mostly useful if there is symmetry in sample sizes even if there are differences in total mRNA across samples.

DE analysis involves analyzing significant differences in the quantitative levels of transcripts or exons among various samples and conditions. Since gene regulatory mechanisms often collate multiple genes at a time or under a single condition, obtaining statistically significant information becomes tricky. This is attributed to quantitative differences in sample size and expression data as a comparatively large RNA-Seq data could be generated for a limited set of samples [94, 95]. Under this scenario, DE analysis essentially requires multivariate statistical methods that include principle component analysis (PCA), canonical correlation analysis (CCA), and nonnegative matrix factorization (NMF) [96]. Thus, statistical software like R and its bioconductor packages find a high utility in performing DE analysis for RNA-Seq data. There are several tools including *Cuffdiff* [81] and bioconductor packages such as *DESeq2* [97], *edgeR* [98], and *Limma* [99] that are frequently used for DE analysis.

All three packages utilize linear models for a comparative correlation between gene expression/transcript abundance with the variables listed by the user. Additionally, these packages need designated design matrix or experimental information containing outcomes and parameters together with a contrast matrix that describes the comparisons to be analyzed. The early steps in the DE analysis essentially require creating the input count table that could be generated as described below.

BAM Files The alignment files in BAM format are converted to SAM files first (*HTSeq*) or can be used directly (*BEDTools*). Both tools produce a count table that could be used for DE analysis.

Individual Count Files Individual count files generated using *HTSeq* can be combined either using UNIX or R commands directly. In addition, *DESeq2* has a separate script for combining the individual files into single count table. In our RNA-Seq pipeline we use a custom R script (https://github.com/grimmlab/BookChapter-RNA-Seq-Analyses/blob/master/Rscripts/DE_deseq_genes_bedtools.R). We first perform a rlog transformations and then a DE analysis using *DESeq2*:

```
Rscript DE_deseq_genes_bedtools.R \
-genecountanalysis/quantification/bedtools-count/
Read_per_features_combined.csv \
-metadata metadata.txt \
-condition condition \
-outputpath analysis/DE/deseq2/
```

Existing Count Table If a pre-existing count table is available, then it can be processed directly by either package.

Subsequently, the lowly expressed transcripts are removed by independent filtering before the count-based DE analysis is performed using either of the packages (*DESeq2, edgeR,* or *Limma*).

DESeq2 is a widely used DE analysis package that utilizes negative binomial generalized linear models for estimating dispersion and fold changes from the distribution extracted from the count table. It mainly functions based on a data frame containing group definitions and other information. *DESeq2* defines relevant groups and prepares a data frame based on the available group information. Following this, different models are utilized for extracting fold changes and distributions. In addition, the count data generated by *DEXSeq* and *Cufflinks* have been used for DE analysis in the proposed RNA-Seq workflow, either directly or by using customized R-scripts.

DEXSeq

```
Rscript DE_dexseq_exon.R \
-exoncountpath analysis/quantification/dexseq-count/ \
-gffFile analysis/quantification/dexseq-count/DEXSEQ_GTF_annotation.gff \
-metadata metadata.txt \
-outputpath analysis/DE/dexseq/
```

Cufflinks

```
cuffdiff \
-o analysis/DE/cuffdiff \
-L vt,dt \
-FDR 0.01 \
-u data/mouse-gencode-version-24/gencode.vM24.annotation.gtf \
-p 8 \
BAMLIST
```

```
Description of parameters:
-o          path to the output folder
-L          lists the labels to be used as "conditions"
-FDR        cutoff for false discovery rate for the DE analysis
-u          path to annotation file
-p          is the number threads
BAMLIST is list of all bam file in comma format
```

Review Question 3

What is differential gene expression?

11.15 Functional Analysis

Another important downstream analysis of RNA-Seq data involves gene set enrichment analysis using functional annotation of differentially expressed (DE) genes or transcripts. This simply means identifying association of DE genes/transcripts with molecular function or with a particular biological process [100]. Functional analysis could be performed in many ways including clustering analysis and gene ontology methods. Clustering analysis usually involves identification of shared promoters or other upstream genomic elements for predicting correlation between gene co-expression and known biological functions [96]. On the other hand, biological ontology method involves annotation of genes to biological functions using graph structures from the Kyoto Encyclopedia of Genes and Genomes (KEGG) or gene ontology (GO) terms [101]. There are several tools available for performing functional analysis such as *GO::Term Finder* [102], *GSEA* [103], and *Clusterprofiler* [104]. The Gene Set Enrichment Analysis (GSEA) utilizes DE data for identifying gene sets or groups of genes that share common biological function, chromosomal location, or regulation. Similarly, the *Clusterprofiler* package is implemented for gene cluster assessment and comparison of biological patterns present in them.

Take Home Message
- Next-generation sequencing (NGS) based transcriptomic analysis provides higher sequence coverage, detection of low abundance and novel transcripts, dynamic changes in mRNA expression levels, and precisely catalogues genetic variants, splice variants, and protein isoforms.
- Several RNA-Seq experimental and analytical protocols are available for analyzing a wide variety of sample types with variable sample qualities.

(continued)

- Choice of the RNA extraction and library preparation methods, NGS platform, quality check, and pre-processing protocols affect the outcomes of RNA-Seq experiments.
- RNA-Seq data processing highly depends on the presence or absence of a reference genome that remains the pivotal determinant of the choice of a reference-based alignment or a *de novo* assembly for a given dataset.
- We provide a best-practice RNA-Seq analysis workflow on GitHub: https://github.com/grimmlab/BookChapter-RNA-Seq-Analyses. This workflow is a stepwise reference-based RNA-Seq analysis example and shall help to get started with a basic RNA-Seq analysis.

Further Reading

- The Biostar Handbook: 2nd Edition.
    https://www.biostarhandbook.com/
- GNU Bash Reference Manual by Chet Ramey, Brian Fox.
    https://www.gnu.org/software/bash/manual/bash.pdf
- An Introduction to Statistical Learning with Applications in R by Gareth James, Daniela Witten, Trevor Hastie, and Robert Tibshirani.
    http://faculty.marshall.usc.edu/gareth-james/ISL/
- Python for Bioinformatics by Sebastian Bassi.
- Next-generation transcriptome assembly. Martin JA, Wang Z. Nat Rev Genet. 2011 Sep 7;12(10):671-82.
- RNA-Seq: a revolutionary tool for transcriptomics. Wang Z, Gerstein M, Snyder M. Nat Rev Genet. 2009 Jan;10(1):57–63.
- Computational and analytical challenges in single-cell transcriptomics. Stegle O, Teichmann SA, Marioni JC. Nat Rev Genet. 2015 Mar;16(3):133–45.
- Genetic Variation and the De Novo Assembly of Human Genomes, Chaisson MJP, Wilson RK, Eichler EE, Nat Rev Genet. 2015 Nov; 16(11):627–40.

Answers to Review Questions

Answer to Questions 1: During the read alignment the first crucial step is to determine the point of origin of the read sequence with respect to the reference genome. For an accurate mapping it is important to consider that reads might originate from either cDNA corresponding to spliced transcripts or from non-spliced transcripts. In the case of spliced transcripts, contiguous read sequences are separated by intervening splice junction (SJ) boundaries and are split into two fragments and assigned separately. Thus, there are two main alignment approaches for aligning reads against a reference genome, that is a spliced unaware and spliced aware alignment. For this purpose, the RNA-Seq reads are typically mapped to either a genome (splice aware) or a transcriptome (splice unaware). Most of the splice unaware alignment tools align DNA against DNA and would have to introduce a long gap in the mapping of a read to span an intron of varying length. Thus, splice unaware aligners are commonly used if reads can be directly mapped against a reference transcriptome or prokaryotic genome. An example of an ungapped aligner is Bowtie, which is an alignment tool based on Burrows–Wheeler Transforms (BWT). Note that this strategy will not support the discovery of novel transcripts. Especially transcriptomes of alternatively spliced organisms, like eukaryotes, are generally suitable for most reference-genome based alignments. As the information on SJs in the reference genome is crucial, most aligners also refer to known annotated SJ databases to confirm the presence or absence of splice sites. Splice-aware aligner like STAR incorporates a three-step alignment procedure to map the reads. In the first step, sequencing reads are aligned to a reference genome. Next, overlapping reads present at each locus are collated into a graph representing all isoforms. In the final step, the graph is resolved, and all isoforms associated with individual genes are identified.

Answer to Questions 2: Gene expression quantification can be done by either counting reads per genes, transcripts, or exons. In the first case, gene expression is quantified by counting the number of reads that align to each gene in the RNA-Seq data. In the second case, quantification based on counting reads per transcript involves precise estimation of transcript abundances followed by an expectation maximization (EM) approach for abundance-based assignments of reads to different transcripts. In the third case, gene expression quantification using exons involves identification and indexing sets of nonoverlapping exonic regions. The choice of the correct quantification depends on the study design and will have a high impact on the outcome of an experiment.

Answer to Question 3: The aim of differential expression analysis is to identify genes with difference in expression patterns under different experimental conditions (gene(s) versus condition(s) and vice versa) for a set of samples.

Acknowledgements We are grateful to Dr. Philipp Torkler (Senior Bioinformatics Scientist, *Exosome Diagnostics, a Bio-Techne brand*, Munich, Germany) for critically reading this text. We thank for correcting our mistakes and suggesting relevant improvements to the original manuscript.

References

1. Crick F. Central dogma of molecular biology. Nature. 1970;227(5258):561–3.
2. de Smith AJ, Walters RG, Froguel P, Blakemore AI. Human genes involved in copy number variation: mechanisms of origin, functional effects and implications for disease. Cytogenet Genome Res. 2008;123(1-4):17–26.
3. Jonkhout N, Tran J, Smith MA, Schonrock N, Mattick JS, Novoa EM. The RNA modification landscape in human disease. RNA. 2017;23(12):1754–69.
4. Martin JA, Wang Z. Next-generation transcriptome assembly. Nat Rev Genet. 2011;12 (10):671–82.

5. Byron SA, Van Keuren-Jensen KR, Engelthaler DM, Carpten JD, Craig DW. Translating RNA sequencing into clinical diagnostics: opportunities and challenges. Nat Rev Genet. 2016;17 (5):257–71.
6. Wang Z, Gerstein M, Snyder M. RNA-Seq: a revolutionary tool for transcriptomics. Nat Rev Genet. 2009;10(1):57–63.
7. Becker-Andre M, Hahlbrock K. Absolute mRNA quantification using the polymerase chain reaction (PCR). A novel approach by a PCR aided transcript titration assay (PATTY). Nucleic Acids Res. 1989;17(22):9437–46.
8. Hoheisel JD. Microarray technology: beyond transcript profiling and genotype analysis. Nat Rev Genet. 2006;7(3):200–10.
9. Bumgarner R. Overview of DNA microarrays: types, applications, and their future. Curr Protoc Mol Biol. 2013;101:22-1.
10. Russo G, Zegar C, Giordano A. Advantages and limitations of microarray technology in human cancer. Oncogene. 2003;22(42):6497–507.
11. Zhao S, Fung-Leung WP, Bittner A, Ngo K, Liu X. Comparison of RNA-Seq and microarray in transcriptome profiling of activated T cells. PLoS One. 2014;9(1):e78644.
12. Schuierer S, Carbone W, Knehr J, Petitjean V, Fernandez A, Sultan M, et al. A comprehensive assessment of RNA-seq protocols for degraded and low-quantity samples. BMC Genomics. 2017;18(1):442.
13. Holik AZ, Law CW, Liu R, Wang Z, Wang W, Ahn J, et al. RNA-seq mixology: designing realistic control experiments to compare protocols and analysis methods. Nucleic Acids Res. 2017;45(5):e30.
14. Kukurba KR, Montgomery SB. RNA sequencing and analysis. Cold Spring Harb Protoc. 2015;2015(11):951–69.
15. Scholes AN, Lewis JA. Comparison of RNA isolation methods on RNA-Seq: implications for differential expression and meta-analyses. BMC Genomics. 2020;21(1):249.
16. Ali N, Rampazzo RCP, Costa ADT, Krieger MA. Current nucleic acid extraction methods and their implications to point-of-care diagnostics. Biomed Res Int. 2017;2017:9306564.
17. Gallego Romero I, Pai AA, Tung J, Gilad Y. RNA-seq: impact of RNA degradation on transcript quantification. BMC Biol. 2014;12:42.
18. Bivens NJ, Zhou M. RNA-Seq library construction methods for transcriptome analysis. Curr Protoc Plant Biol. 2016;1(1):197–215.
19. Wang L, Felts SJ, Van Keulen VP, Pease LR, Zhang Y. Exploring the effect of library preparation on RNA sequencing experiments. Genomics. 2019;111(6):1752–9.
20. Liu L, Li Y, Li S, Hu N, He Y, Pong R, et al. Comparison of next-generation sequencing systems. J Biomed Biotechnol. 2012;2012:251364.
21. Kovaka S, Zimin AV, Pertea GM, Razaghi R, Salzberg SL, Pertea M. Transcriptome assembly from long-read RNA-seq alignments with StringTie2. Genome Biol. 2019;20(1):278.
22. Hwang B, Lee JH, Bang D. Single-cell RNA sequencing technologies and bioinformatics pipelines. Exp Mol Med. 2018;50(8):96.
23. Quail MA, Smith M, Coupland P, Otto TD, Harris SR, Connor TR, et al. A tale of three next generation sequencing platforms: comparison of Ion Torrent, Pacific Biosciences and Illumina MiSeq sequencers. BMC Genomics. 2012;13:341.
24. Mohorianu I, Bretman A, Smith DT, Fowler EK, Dalmay T, Chapman T. Comparison of alternative approaches for analysing multi-level RNA-seq data. PLoS One. 2017;12(8): e0182694.
25. Andrews S. FASTQC. A quality control tool for high throughput sequence data. 2010.
26. Planet E, Attolini CS, Reina O, Flores O, Rossell D. htSeqTools: high-throughput sequencing quality control, processing and visualization in R. Bioinformatics. 2012;28(4):589–90.

27. Lassmann T, Hayashizaki Y, Daub CO. SAMStat: monitoring biases in next generation sequencing data. Bioinformatics. 2011;27(1):130–1.
28. Martin M. Cutadapt removes adapter sequences from high-throughput sequencing reads. EMBnet J. 2011;17(1):3.
29. Bolger AM, Lohse M, Usadel B. Trimmomatic: a flexible trimmer for Illumina sequence data. Bioinformatics. 2014;30(15):2114–20.
30. Leinonen R, Sugawara H, Shumway M. International nucleotide sequence database C. The sequence read archive. Nucleic Acids Res. 2011;39(Database issue):D19–21.
31. Armstrong J, Fiddes IT, Diekhans M, Paten B. Whole-genome alignment and comparative annotation. Annu Rev Anim Biosci. 2019;7:41–64.
32. Conesa A, Madrigal P, Tarazona S, Gomez-Cabrero D, Cervera A, McPherson A, et al. A survey of best practices for RNA-seq data analysis. Genome Biol. 2016;17:13.
33. Lee RS, Behr MA. Does choice matter? Reference-based alignment for molecular epidemiology of tuberculosis. J Clin Microbiol. 2016;54(7):1891–5.
34. Sherman RM, Salzberg SL. Pan-genomics in the human genome era. Nat Rev Genet. 2020;21(4):243–54.
35. Engstrom PG, Steijger T, Sipos B, Grant GR, Kahles A, Ratsch G, et al. Systematic evaluation of spliced alignment programs for RNA-seq data. Nat Methods. 2013;10(12):1185–91.
36. Trapnell C, Salzberg SL. How to map billions of short reads onto genomes. Nat Biotechnol. 2009;27(5):455–7.
37. Baruzzo G, Hayer KE, Kim EJ, Di Camillo B, FitzGerald GA, Grant GR. Simulation-based comprehensive benchmarking of RNA-seq aligners. Nat Methods. 2017;14(2):135–9.
38. Bushnell B, editor. BBMap: a fast, accurate, splice-aware aligner. Berkeley: Lawrence Berkeley National Lab.(LBNL); 2014.
39. Dobin A, Davis CA, Schlesinger F, Drenkow J, Zaleski C, Jha S, et al. STAR: ultrafast universal RNA-seq aligner. Bioinformatics. 2012;29(1):15–21.
40. Wu TD, Watanabe CK. GMAP: a genomic mapping and alignment program for mRNA and EST sequences. Bioinformatics. 2005;21(9):1859–75.
41. Kim D, Pertea G, Trapnell C, Pimentel H, Kelley R, Salzberg SL. TopHat2: accurate alignment of transcriptomes in the presence of insertions, deletions and gene fusions. Genome Biol. 2013;14(4):R36.
42. Benjamin AM, Nichols M, Burke TW, Ginsburg GS, Lucas JE. Comparing reference-based RNA-Seq mapping methods for non-human primate data. BMC Genomics. 2014;15:570.
43. Zakeri M, Srivastava A, Almodaresi F, Patro R. Improved data-driven likelihood factorizations for transcript abundance estimation. Bioinformatics. 2017;33(14):i142–i51.
44. Wu DC, Yao J, Ho KS, Lambowitz AM, Wilke CO. Limitations of alignment-free tools in total RNA-seq quantification. BMC Genomics. 2018;19(1):510.
45. Langmead B, Salzberg SL. Fast gapped-read alignment with Bowtie 2. Nat Methods. 2012;9(4):357–9.
46. Li H, Ruan J, Durbin R. Mapping short DNA sequencing reads and calling variants using mapping quality scores. Genome Res. 2008;18(11):1851–8.
47. Frankish A, Diekhans M, Ferreira A-M, Johnson R, Jungreis I, Loveland J, et al. GENCODE reference annotation for the human and mouse genomes. Nucleic Acids Res. 2018;47(D1):D766–D73.
48. Schaarschmidt S, Fischer A, Zuther E, Hincha DK. Evaluation of seven different RNA-Seq alignment tools based on experimental data from the model plant Arabidopsis thaliana. Int J Mol Sci. 2020;21(5):1720.
49. Holzer M, Marz M. De novo transcriptome assembly: a comprehensive cross-species comparison of short-read RNA-Seq assemblers. Gigascience. 2019;8(5):giz039.

50. Moreton J, Izquierdo A, Emes RD. Assembly, assessment, and availability of De novo generated eukaryotic transcriptomes. Front Genet. 2015;6:361.
51. Gopinath GR, Cinar HN, Murphy HR, Durigan M, Almeria M, Tall BD, et al. A hybrid reference-guided de novo assembly approach for generating Cyclospora mitochondrion genomes. Gut Pathog. 2018;10:15.
52. Lischer HEL, Shimizu KK. Reference-guided de novo assembly approach improves genome reconstruction for related species. BMC Bioinform. 2017;18(1):474.
53. Martin J, Bruno VM, Fang Z, Meng X, Blow M, Zhang T, et al. Rnnotator: an automated de novo transcriptome assembly pipeline from stranded RNA-Seq reads. BMC Genomics. 2010;11:663.
54. Robertson G, Schein J, Chiu R, Corbett R, Field M, Jackman SD, et al. De novo assembly and analysis of RNA-seq data. Nat Methods. 2010;7(11):909–12.
55. Grabherr MG, Haas BJ, Yassour M, Levin JZ, Thompson DA, Amit I, et al. Full-length transcriptome assembly from RNA-Seq data without a reference genome. Nat Biotechnol. 2011;29(7):644–52.
56. Compeau PE, Pevzner PA, Tesler G. How to apply de Bruijn graphs to genome assembly. Nat Biotechnol. 2011;29(11):987–91.
57. Zerbino DR, Birney E. Velvet: algorithms for de novo short read assembly using de Bruijn graphs. Genome Res. 2008;18(5):821–9.
58. Schulz MH, Zerbino DR, Vingron M, Birney E. Oases: robust de novo RNA-seq assembly across the dynamic range of expression levels. Bioinformatics. 2012;28(8):1086–92.
59. O'Neil ST, Emrich SJ. Assessing De Novo transcriptome assembly metrics for consistency and utility. BMC Genomics. 2013;14:465.
60. Moreno-Santillan DD, Machain-Williams C, Hernandez-Montes G, Ortega J. De Novo transcriptome assembly and functional annotation in five species of bats. Sci Rep. 2019;9 (1):6222.
61. Evangelistella C, Valentini A, Ludovisi R, Firrincieli A, Fabbrini F, Scalabrin S, et al. De novo assembly, functional annotation, and analysis of the giant reed (Arundo donax L.) leaf transcriptome provide tools for the development of a biofuel feedstock. Biotechnol Biofuels. 2017;10:138.
62. UniProt C. The universal protein resource (UniProt). Nucleic Acids Res. 2008;36(Database issue):D190–5.
63. Finn RD, Bateman A, Clements J, Coggill P, Eberhardt RY, Eddy SR, et al. Pfam: the protein families database. Nucleic Acids Res. 2014;42(Database issue):D222–30.
64. Lagesen K, Hallin P, Rodland EA, Staerfeldt HH, Rognes T, Ussery DW. RNAmmer: consistent and rapid annotation of ribosomal RNA genes. Nucleic Acids Res. 2007;35(9):3100–8.
65. Almagro Armenteros JJ, Tsirigos KD, Sonderby CK, Petersen TN, Winther O, Brunak S, et al. SignalP 5.0 improves signal peptide predictions using deep neural networks. Nat Biotechnol. 2019;37(4):420–3.
66. Conesa A, Gotz S, Garcia-Gomez JM, Terol J, Talon M, Robles M. Blast2GO: a universal tool for annotation, visualization and analysis in functional genomics research. Bioinformatics. 2005;21(18):3674–6.
67. Kanehisa M, Goto S. KEGG: kyoto encyclopedia of genes and genomes. Nucleic Acids Res. 2000;28(1):27–30.
68. Li H, Handsaker B, Wysoker A, Fennell T, Ruan J, Homer N, et al. The sequence alignment/map format and SAMtools. Bioinformatics. 2009;25(16):2078–9.
69. Barnett DW, Garrison EK, Quinlan AR, Stromberg MP, Marth GT. BamTools: a C++ API and toolkit for analyzing and managing BAM files. Bioinformatics. 2011;27(12):1691–2.

70. Tarasov A, Vilella AJ, Cuppen E, Nijman IJ, Prins P. Sambamba: fast processing of NGS alignment formats. Bioinformatics. 2015;31(12):2032–4.
71. Tischler G, Leonard S. biobambam: tools for read pair collation based algorithms on BAM files. Source Code Biol Med. 2014;9(1):13.
72. Wang L, Wang S, Li W. RSeQC: quality control of RNA-seq experiments. Bioinformatics. 2012;28(16):2184–5.
73. Buels R, Yao E, Diesh CM, Hayes RD, Munoz-Torres M, Helt G, et al. JBrowse: a dynamic web platform for genome visualization and analysis. Genome Biol. 2016;17:66.
74. Robinson JT, Thorvaldsdottir H, Winckler W, Guttman M, Lander ES, Getz G, et al. Integrative genomics viewer. Nat Biotechnol. 2011;29(1):24–6.
75. Kent WJ, Sugnet CW, Furey TS, Roskin KM, Pringle TH, Zahler AM, et al. The human genome browser at UCSC. Genome Res. 2002;12(6):996–1006.
76. Kallio MA, Tuimala JT, Hupponen T, Klemela P, Gentile M, Scheinin I, et al. Chipster: user-friendly analysis software for microarray and other high-throughput data. BMC Genomics. 2011;12:507.
77. Jin H, Wan YW, Liu Z. Comprehensive evaluation of RNA-seq quantification methods for linearity. BMC Bioinform. 2017;18(Suppl 4):117.
78. Teng M, Love MI, Davis CA, Djebali S, Dobin A, Graveley BR, et al. A benchmark for RNA-seq quantification pipelines. Genome Biol. 2016;17:74.
79. Anders S, Pyl PT, Huber W. HTSeq–a Python framework to work with high-throughput sequencing data. Bioinformatics. 2015;31(2):166–9.
80. Quinlan AR, Hall IM. BEDTools: a flexible suite of utilities for comparing genomic features. Bioinformatics. 2010;26(6):841–2.
81. Trapnell C, Williams BA, Pertea G, Mortazavi A, Kwan G, van Baren MJ, et al. Transcript assembly and quantification by RNA-Seq reveals unannotated transcripts and isoform switching during cell differentiation. Nat Biotechnol. 2010;28(5):511–5.
82. Roberts A, Pachter L. Streaming fragment assignment for real-time analysis of sequencing experiments. Nat Methods. 2013;10(1):71–3.
83. Anders S, Reyes A, Huber W. Detecting differential usage of exons from RNA-seq data. Genome Res. 2012;22(10):2008–17.
84. Wang T, Li B, Nelson CE, Nabavi S. Comparative analysis of differential gene expression analysis tools for single-cell RNA sequencing data. BMC Bioinform. 2019;20(1):40.
85. Lamarre S, Frasse P, Zouine M, Labourdette D, Sainderichin E, Hu G, et al. Optimization of an RNA-Seq differential gene expression analysis depending on biological replicate number and library size. Front Plant Sci. 2018;9:108.
86. Conesa A, Madrigal P, Tarazona S, Gomez-Cabrero D, Cervera A, McPherson A, et al. A survey of best practices for RNA-seq data analysis. Genome Biol. 2016;17:13.
87. Gonzalez E, Joly S. Impact of RNA-seq attributes on false positive rates in differential expression analysis of de novo assembled transcriptomes. BMC Res Notes. 2013;6:503.
88. Mandelboum S, Manber Z, Elroy-Stein O, Elkon R. Recurrent functional misinterpretation of RNA-seq data caused by sample-specific gene length bias. PLoS Biol. 2019;17(11):e3000481.
89. Wang Z, Gerstein M, Snyder M. RNA-Seq: a revolutionary tool for transcriptomics. Nat Rev Genet. 2009;10(1):57–63.
90. Li X, Cooper NGF, O'Toole TE, Rouchka EC. Choice of library size normalization and statistical methods for differential gene expression analysis in balanced two-group comparisons for RNA-seq studies. BMC Genomics. 2020;21(1):75.
91. Bullard JH, Purdom E, Hansen KD, Dudoit S. Evaluation of statistical methods for normalization and differential expression in mRNA-Seq experiments. BMC Bioinform. 2010;11:94.

92. Evans C, Hardin J, Stoebel DM. Selecting between-sample RNA-Seq normalization methods from the perspective of their assumptions. Brief Bioinform. 2018;19(5):776–92.

93. Li P, Piao Y, Shon HS, Ryu KH. Comparing the normalization methods for the differential analysis of Illumina high-throughput RNA-Seq data. BMC Bioinform. 2015;16:347.

94. Costa-Silva J, Domingues D, Lopes FM. RNA-Seq differential expression analysis: an extended review and a software tool. PLoS One. 2017;12(12):e0190152.

95. Soneson C, Delorenzi M. A comparison of methods for differential expression analysis of RNA-seq data. BMC Bioinform. 2013;14:91.

96. Zhou Y, Zhu J, Tong T, Wang J, Lin B, Zhang J. A statistical normalization method and differential expression analysis for RNA-seq data between different species. BMC Bioinform. 2019;20(1):163.

97. Love MI, Huber W, Anders S. Moderated estimation of fold change and dispersion for RNA-seq data with DESeq2. Genome Biol. 2014;15(12):550.

98. Robinson MD, McCarthy DJ, Smyth GK. edgeR: a Bioconductor package for differential expression analysis of digital gene expression data. Bioinformatics. 2010;26(1):139–40.

99. Ritchie ME, Phipson B, Wu D, Hu Y, Law CW, Shi W, et al. limma powers differential expression analyses for RNA-sequencing and microarray studies. Nucleic Acids Res. 2015;43 (7):e47.

100. Huang Y, Pan J, Chen D, Zheng J, Qiu F, Li F, et al. Identification and functional analysis of differentially expressed genes in poorly differentiated hepatocellular carcinoma using RNA-seq. Oncotarget. 2017;8(22):35973–83.

101. Martin D, Brun C, Remy E, Mouren P, Thieffry D, Jacq B. GOToolBox: functional analysis of gene datasets based on Gene ontology. Genome Biol. 2004;5(12):R101.

102. Boyle EI, Weng S, Gollub J, Jin H, Botstein D, Cherry JM, et al. GO::TermFinder–open source software for accessing gene ontology information and finding significantly enriched Gene Ontology terms associated with a list of genes. Bioinformatics. 2004;20(18):3710–5.

103. Subramanian A, Tamayo P, Mootha VK, Mukherjee S, Ebert BL, Gillette MA, et al. Gene set enrichment analysis: a knowledge-based approach for interpreting genome-wide expression profiles. Proc Natl Acad Sci U S A. 2005;102(43):15545–50.

104. Yu G, Wang LG, Han Y, He QY. clusterProfiler: an R package for comparing biological themes among gene clusters. OMICS. 2012;16(5):284–7.

Design and Analysis of Epigenetics and ChIP-Sequencing Data

12

Melanie Kappelmann-Fenzl

Contents

What You Will Learn in This Chapter
In this chapter the theoretical background, the experimental requirements, and some ways to evaluate Chromatin Immunoprecipitation followed by NGS (ChIP-Seq) are represented and explained using practical examples. You will learn about the main differences between sequencing DNA regions with certain histone modifications or transcription factor binding sites. Moreover, we will introduce a software tool *HOMER*, which offers a variety of (epigenetic) sequencing data analysis options.

(continued)

M. Kappelmann-Fenzl (✉)
Deggendorf Institute of Technology, Deggendorf, Germany

Institute of Biochemistry (Emil-Fischer Center), Friedrich-Alexander University Erlangen-Nürnberg, Erlangen, Germany
e-mail: melanie.kappelmann-fenzl@th-deg.de

© Springer Nature Switzerland AG 2021
M. Kappelmann-Fenzl (ed.), *Next Generation Sequencing and Data Analysis*, Learning
Materials in Biosciences, https://doi.org/10.1007/978-3-030-62490-3_12

Herefore, the most important scripts, commands, and options and their purpose are illustrated in this chapter. After you have worked through this chapter you will understand the impact of epigenetic sequencing approaches and you will be able to perform the ChIP-Seq data analysis workflow—from receiving your raw data after sequencing to motif discovery in your identified ChIP-Seq peaks/regions.

12.1 Introduction

Epigenetic sequencing approaches allow to study heritable or acquired changes in gene activity caused by mechanisms other than DNA sequence changes. Epigenetic analysis research can involve studying alterations in DNA methylation, DNA–protein interactions, chromatin accessibility, histone modifications, and more, on a genome-wide scale. In this textbook we focus on analyzing ChIP-Seq data based on DNA–protein interaction of transcription factors or (modified) histones. However, the sequencing data analysis workflow is, with minor differences, similar for all approaches. The main aim of ChIP-Seq approaches is to identify genetic regulatory networks (GRNs) to determine transcriptionally active genes in any cell type of interest. Genes are transcribed by RNA Polymerase II, but binding by specific transcription factors is required to initialize this process. The following simplified illustration depicts the phenomenon of gene regulation by a specific regulatory protein (transcription factor, TF), without which transcription does not occur (Fig. 12.1).

Thus, ChIP-Seq data provide insights into regulation events by identification of transcription factor binding sites, so-called binding motifs, within a promoter sequence or other regulatory sequences (enhancer/silencer). Moreover, ChIP-Seq data can be used to track histone modifications across the genome, and narrow in on chromatin structure and function. Next Generation Sequencing reads from ChIP-Seq experiments can be evaluated by different software tools in different ways. This textbook describes an open source software called *HOMER* [1] and Bioconductor packages in R to analyze ChIP-Seq data.

12.2 DNA Quality and ChIP-Seq Library Preparation

For successful ChIP-Seq approaches, one must generate high-quality ChIP-DNA templates to obtain the best sequencing outcomes. ChIP-Seq experiments typically begin with the formaldehyde cross-linking of protein–DNA complexes in cells or tissue. The chromatin is then extracted and fragmented, either through enzymatic digestion or sonication, and DNA–protein fragments are immunoprecipitated with target-specific antibodies (the target

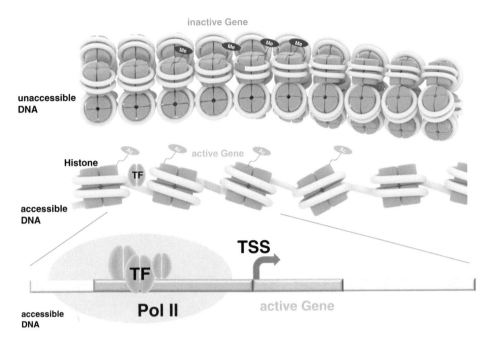

Fig. 12.1 Epigenetic modifications effect chromatin structure and thus transcriptional activation. Histones are proteins around which DNA winds for compaction and gene regulation. DNA methylation (not shown) and chemical modification of histone tails (acetylation= Ac, or methylation=Mc) alter the spacing of nucleosomes and change expression of associated genes. Transcription factor binding in promoter, silencer, or enhancer regions of the DNA also effects gene expression. Active promoter regions are accessible for common (gray) and specific (TF; blue) transcription factors, which are then responsible for recruiting Polymerase II (Pol II) to initiate transcription and perform RNA synthesis starting at the TSS (Transcription Start Site) of an active gene. (© Melanie Kappelmann-Fenzl)

is the respective TF or histone). Generating reliable ChIP-Seq data depends on using antibodies that have been validated for target specificity and acceptable signal-to-noise ratios to perform the ChIP experiment (Fig. 12.2) [2].

The amount of ChIP-DNA to use when creating a DNA library is influenced by factors like the amount of DNA obtained from the actual ChIP, the desired library yield, and the limits of PCR amplification required to minimize duplicate sequencing reads. A typical histone ChIP experiment using 10 μg of input chromatin DNA per immunoprecipitation yields approximately 100–1000 ng of ChIP-DNA. In comparison, a transcription factor or cofactor ChIP experiment yields approximately 5–25 ng of ChIP-DNA. However, the conventional single-step cross-linking technique does not preserve all protein–DNA interactions, especially for transcription factors or for coactivator interactions. Thus, for these cases it is recommended to perform an additional DNA–protein fixation step using DSG (Disuccinimidyl glutarate) [3]. Finally, the quality and quantity of ChIP-DNA can be assessed by either Agilent Bioanalyzer or TapeStation systems (https://www.agilent. com/cs/library/catalogs/public/Catalog-bioanalyzer-tapestation-systems-sw-consumables-

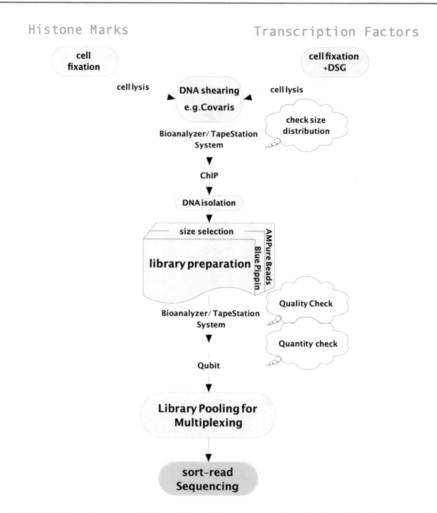

Fig. 12.2 A schematic representation of the various steps involved in ChIP of either Histone Marks or Transcription Factors followed by ChIP-Seq library preparation and sequencing

5994-0249EN_agilent.pdf) (Fig. 12.3). It is recommended using 50 ng of histone ChIP-DNA and 5 ng of transcription factor and cofactor ChIP-DNA for library construction [4, 5].

The workflow to then achieve a high-quality ChIP-Seq library (see Chap. 3) is provided in the Appendix section (Table 13.1).

12.3   Quality Check (QC) and Sequencing Pre-processing

After sequencing your ChIP-Seq libraries you can start over with the bioinformatical data analysis part. A whole example workflow is illustrated in the Flow-Chart (Fig. 12.4).

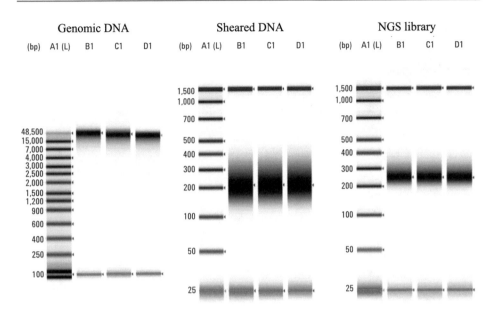

Fig. 12.3 Gel images of the QC steps obtained from the Agilent 2200 TapeStation system for genomic DNA, purified, sheared DNA, and a NGS library as analyzed on the Genomic DNA, D1000, and High Sensitivity D1000 ScreenTape, respectively. The gel images show high-quality samples. (source: modified according to https://www.agilent.com/cs/library/applications/5991-3654EN.pdf)

As already described in Chap. 7, you first have to transform your ChIP-Seq data (unaligned *.bam*) into *.fastq* file format. First, change your working directory to the folder where your raw sequencing data are located:

```
cd /path/to/rawData
```

In general, ChIP-Seq is performed by sequencing 50 bp single end, thus the command to generate *.fastq* files is:

```
bedtools bamtofastq -I *.bam -fq *.fastq
```

In a next step, quality control can be performed on the generated *.fastq* files by the *FastQC* tool (https://www.bioinformatics.babraham.ac.uk/projects/fastqc/):

```
fastqc *_1.fastq *_2.fastq *_3.fastq
```

The output of `fastqc` is an html file and can be opened using any Internet browser. For a detailed description of *FastQC Report* see Sect. 7.3.1.

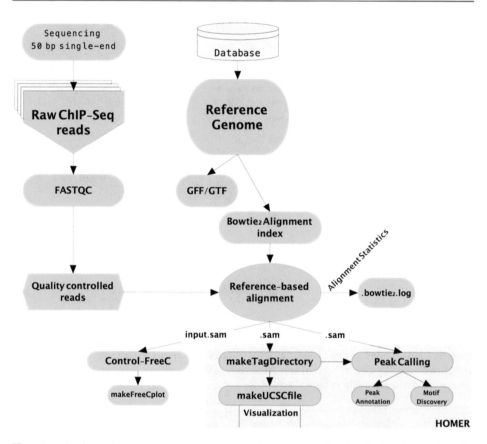

Fig. 12.4 A schematic representation of the various steps and tools involved in the described ChIP-Seq analysis strategy

After quality check, the sequencing data can be mapped to the reference genome. In terms of ChIP-Seq data a fast and sensitive alignment tool like Bowtie2 [6] is a suitable choice (see Sect. 9.2.3.3).

One possible mapping command is:

```
bowtie2 -x GRCh38 -p 23 -un -name /path/to/outputfile //nameExtension /path/
to/.fastq
```

```
Description of parameters:
-x           Index filename prefix (minus trailing .X.bt2)
-p           number of alignment threads to launch
-un          will output unaligned reads
-name        path/outputfile name w/o extension
```

This mapping command will output a *.sam* file (see Sect. 7.2.4) for each *.fastq* file in the defined output directory and **.bowtie2.unaligned.fq* as well as **.bowtie2.log* files (summary of mapping progress) in the input directory.

Next, we place all relevant information about the experiment into a so-called tag directory, which is essentially a directory on your computer that contains several files describing your experiment. Therefore, we use the `makeTagDirectory` script of the *HOMER* software tool, which creates a platform-independent "tag directory" for later analysis. The input file for making a "tag directory" is the output file of your mapping process **.sam*.

```
makeTagDirectory /path/to/tagDirectory // nameExtension path/to/
alignment/file/*.sam -genome XY -checkGC
```

```
Description of the used GC bias options:
-genome      genome version (e.g. hg38)
-checkGC     check sequence bias, requires "-genome"
```

To save disk space it is recommended to gzip the *.sam* files:

```
gzip *.sam
```

To create a bedgraph file or bigwig file for visualization using the UCSC Genome or IGV Browser use `makeUCSCfile` and define the output file format.

```
makeUCSCfile /path/to/tagDirectory // nameextension -fsize 1e20 -bigWig /
path/to/GenomeIndices/ /.chromosome.sizes -o /path/to/USCS/tracks/.ucsc.
bigWig -name nameExtension ;
```

```
Description of parameters:
-fsize     Size of file, when gzipped, default: 1e10, i.e. no reduction
-bigWig    <chrom.size file> (creates a full resolution bigWig file and track
           line file). This requires bedGraphToBigWig to be available in your
           executable path.
-name      Name of UCSC track, default: auto generated
```

The program works by approximating the ChIP-fragment density, which is defined as the total number of overlapping fragments at each position in the genome. A detailed description of the different command line options for `makeUCSCfile` can be found here (http://homer. ucsd.edu/homer/ngs/ucsc.html). The visualization of the ChIP-Seq data can provide information on whether there are specific, defined peaks in the data or regions of continuous coverage (histone marks), and whether the reads are distributed over all expected chromosomes. In addition, you can evaluate whether the pattern matches the experiment by looking for specific different patterns:

- TFs: enrichment of reads near the TSS and distal regulatory elements.
- H3K4me3—enrichment of reads near TSS.
- H3K4me1/2, H3/H4ac, DNase—enrichment of reads near TSS and distal regulatory elements.
- H3K36me3—enrichment of reads across gene bodies.
- H3K27me3—enrichment of reads near CpG Islands of inactive genes.
- H3K9me3—enrichment of reads across broad domains and repeat elements.

12.4   Copy Number Variation (CNV) of Input Samples

ChIP-Seq experiments generally show an enrichment of reads in regulatory regions of the genome. Thus, it is important to also sequence a non-chipped genomic DNA of the same sample (Input) to control if the ChIP was successful. The Input sample can be further used to analyze and visualize possible copy number variation (CNV) of the Input sample. This can be important in terms of deletions or duplications of whole chromosomes or parts of a chromosome, which would lead to a false discovery of enriched reads in regions with alterations in copy numbers in the further analysis workflow. Consequently, no or low peaks (enrichment of reads) should be detected in the Input samples compared to the chipped once.

Control-FREEC [7] is a tool we have already installed via `conda` conda for detection of CNV and allelic imbalances (LOH) in NGS data (Chap. 7). It automatically computes, normalizes, segments copy number and beta allele frequency (BAF) profiles. Then it calls CNV and LOH. For whole genome sequencing data analysis, like our Input sample, the program can also use mappability data (files created by GEM (https://sourceforge.net/projects/gemlibrary/files/gem-library/)) [7]. To be able to run Control-FreeC you have to create a FreeC directory in your GenomeIndices folder with all *.fa* files of the reference genome as well as a file with chromosome sizes. Moreover, the mappability file should also be stored here (or elsewhere, but you should remember where). A detailed description of the usage of Control-FreeC can be found on the following website: http://boevalab.inf.ethz.ch/FREEC/tutorial.html. Moreover, the calculation of significance of Control-FreeC predictions and the visualization of Control-FreeC´s output using R are described. The required scripts are stored on GitHub (https://github.com/BoevaLab/FREEC).

12.5   Peak/Region Calling

In terms of ChIP-Seq reads, finding peaks/regions is one of the central goals and the same basic principles apply as for other types of sequencing. In terms of a transcription factor ChIP-Seq experiment one speaks of identifying "peaks," in terms of histone modifications or methylated DNA of "regions." Defining peaks/regions means to identify locations in the genome where we find more sequencing reads than we would expect to find by chance. There are number of different methods for identifying peaks/regions from ChIP-Seq experiments. You can use any peak calling algorithm (http://seabass.mpipz.mpg.de/encode/encodeTools.html) and it is not required that you use *HOMER* for peak finding to use the rest of the tools included in *HOMER*.

First you have to identify the ChIP-Seq peaks/regions and create a so-called position file. An example code for identifying regions of a histone mark ChIP-Seq experiment is depicted below.

```
findPeaks /path/to/ChIPseq/tagDir/GRCh38/SampleName -i /path/to/ChIPseq/
tagDir/GRCh38/InputSampleName -region -size 250 -L 0 -F 5 -minDist 350 –fdr
0.000001 -o /path/to/ChIPseq/analysis/PeakFiles/peaks_SampleName_FDR1e-
6.txt
```

```
Description of parameters:
-i <input tag directory>   Experiment to use as IgG/Input/Control
-region                    extends start/stop coordinates to cover full region
                           considered "enriched". [Alternatively to -region:
                           -style histone (histone modification ChIP-Seq,
                           region based, uses -region -size 500 -L 0, regions.
                           txt)]
-size <#>                  Peak size, default: auto
-L <#>                     fold enrichment over local tag count, default: 4.0
-F <#>                     fold enrichment over input tag count, default: 4.0
-minDist <#>               minimum distance between peaks, default: peak
                           size x2
-fdr <#>                   False discovery rate, default = 0.001
-o <filename|auto>         file name for to output peaks, default: stdout
```

Consequently, the tags of your previously created tag directory of each sample are normalized to the Input and defined as a peak by specific option settings. In terms of a transcription factor ChIP-Seq experiment you would define the setting differently than for histone marks:

```
findPeaks /path/to/ChIPseq/tagDir/GRCh38/SampleName -i /path/to/ChIPseq/
tagDir/GRCh38/InputSampleName -style factor -tbp 1 -fdr 0.000001 -o /path/
to/ChIPseq/analysis/PeakFiles/peaks_SampleName_FDR1e-6.txt
```

```
Description of parameters:
-style    <factor> (transcription factor ChIP-Seq, uses -center, output:
peaks.txt, default)
-tbp      <#> (Maximum tags per bp to count, 0 = no limit, default: auto)
```

In the further course of this textbook we will use for both scenarios "peaks" and "regions" on the name "peaks."

12.6 Further ChIP-Seq Analysis

After peak identification you can merge different peak files to find common/overlapping peaks in different samples and visualize the analysis results via Venn diagrams. A Venn diagram uses overlapping circles or other shapes to illustrate the logical relationships between two or more sets of items (Fig. 12.5). Often, they serve to graphically organize things, highlighting how the items are similar and different. To create a Venn diagram, you can use multiple online tools or R packages.

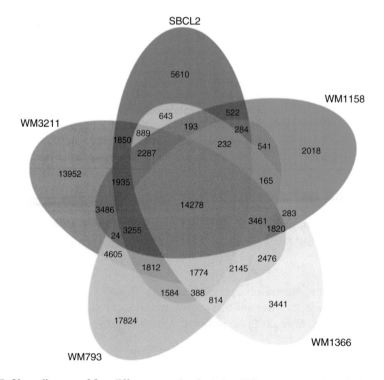

Fig. 12.5 Venn diagram of five different samples depicting differences as well as similarities of the samples. The illustrated diagram was created by using *VennDiagram-package* in R, which can be used to create high-resolution and highly-customizable Venn and Euler plots

```
mergePeaks -d 100 /path/to/ChIPseq/analysis/PeakFiles/
peaks_SampleName1_FDR1e-6.txt /path/to/ChIPseq/analysis/PeakFiles/
peaks_SampleName2_FDR1e-6.txt > /path/to/ChIPseq/analysis/PeakFiles/
merge_SampleName1_SampleName2.txt -venn /path/to/ChIPseq/analysis/Venn/
merge_SampleName1_SampleName2.venn.txt
```

```
Description of parameters:
-d              <#|given> (Maximum distance between peak centers to merge,
                default: given). Use "-d given" when features have vastly
                different sizes (i.e. peaks vs. introns).
-venn           <filename> (output venn diagram numbers to file, default: to
                stderr)
-prefix         <filename> (Generates separate files for overlapping
                and unique peaks)]
```

An example script to create a Venn diagram with five different samples with the output file of `mergePeaks` depicted above can be found here (https://github.com/mkappelmann/ ChIP-Seq-H3K27ac/blob/master/draw.quintuple.venn.R).

Further, you can use *BedTools* utilities [8, 9], which are a Swiss-army knife of tools for a wide-range of genomic analysis tasks. For example, *BedTools* allows one to intersect, merge, count, and complement genomic intervals from multiple files in widely-used genomic file formats such as BAM, BED, GFF/GTF, VCF. For further *BedTools* analysis, the `mergePeaks` output peak file has to be converted into a *.bed* file:

```
pos2bed.pl peakfile.txt > peakfile.bed
#convert .bed into a peak file
bed2pos.pl peakfile.bed > peakfile.txt
```

If you are interested in identifying differences between samples rather than identity, you can use `getDifferentialPeaks`. This command extracts tags near each peak from the tag directories and counts them, by outputting peaks with significantly different tag densities.

Importantly, annotation of peaks is helpful to associate peaks with nearby genes. The basic annotation includes the distance to the next transcription start side (TSS), as well as some other genome annotations like: transcription termination site (TTS), CDS (from coding sequence) exons, 5'-UTR (untranslated region) exons, 3'-UTR exons, CpG islands, repeats, introns, and intergenic. The `annotatePeaks.pl` program also enables you to perform Gene Ontology Analysis, genomic feature association analysis, merge peak files with gene expression data (RNA-Seq) using the `-gene` option, calculate ChIP-Seq Tag densities from different experiments, and find motif occurrences in peaks.

```
annotatePeaks.pl /path/to/ChIPseq/analysis/PeakFiles/
merge_SampleName1_SampleName2.txt hg38 -size 200 -d /path/to/ChIPseq/
tagDir/GRCh38/SampleName1 /path/to/ChIPseq/tagDir/GRCh38/SampleName2 > /
path/to/ChIPseq/analysis/ merge_SampleName1_SampleName2.anno.txt
```

```
Description of parameters:
-size <#>               Peak size [from center of peak
-d <tag directory>      list of experiment directories to show tag counts for
[-gene <data file> ... Adds additional data to result based on the closest
            gene. This is useful for adding gene expression data. The file
            must have a header, and the first column must be a GeneID,
            Accession number, etc. If the peak cannot be mapped to data in the
            file then the entry will be left empty.]
[-hist       -hist <bin size in bp> (i.e 1, 2, 5, 10, 20, 50, 100 etc.). The -hist
            option can be used to generate histograms of position dependent
            features relative to the center of peaks. -ghist (outputs
            profiles for each gene, for peak shape clustering)]
```

One more advanced possibility to use the `annotatePeaks.pl` script is to center all identified peaks /regions relative to the TSS and visualize the results by creating a heatmap:

```
annotatePeaks.pl tss hg38 -size 5000 -hist 50 -ghist -d /path/to/ChIPseq/
tagDir/GRCh38/SampleName -cpu 18 > /path/to/PeakFiles/*_ghist.txt
```

The heatmap can then be generated in R by using the *pheatmap* library:

```
library (pheatmap)

setwd ("/path/to/PeakFiles")
ghist<- read.delim ("*_ghist.txt ", header=TRUE)
head (ghist)
ghist$Gene
m1<- as.matrix( ghist [,2:ncol (ghist)])
rownames (m1) <- ghist$Gene
m1<- m1+1
range (m1)
bk = unique (c (seq (1,14.9, length=100),seq (15,505, length=100)))

hmcols<- colorRampPalette (c ("white","blue")) (length (bk)-1)
pheatmap (m1, color=hmcols, breaks=bk, cluster_rows=FALSE,
cluster_cols=FALSE, legend=FALSE, show_rownames=FALSE,
show_colnames=FALSE)
```

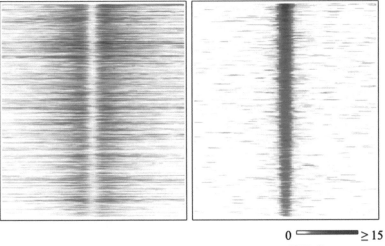

0 ▬▬▬▬▬▬▬ ≥ 15
ChIP-Seq tag count

Fig. 12.6 Heatmaps of identified peaks/ regions centered to TSS. The blue coloring is defined by a tag count of fifteen or more tags of a 100bp length (legend of the ChIP-Seq tag count). These example heatmaps show the distribution of histone modification (left) and TFBSs sequencing tags (right), respectively (modified according to Kappelmann-Fenzl et al. 2019).

R will output a plot depicted in Fig. 12.6 [10]. The TSSs are indicated as a bright line right in the middle of the plot. The identified CHIP-Seq peaks/regions are indicated in darker blue, showing the distribution of sequenced regions around the TSS of each gene (one row in the heatmap) of the genome. As defined in the R script more than 15 sequencing reads (tags) are depicted in blue. Centering can also be performed on any other genome annotation, as well as on defined transcription factor binding sites (TFBS). The latter is performed to identify potential cofactors of TFs.

Furthermore, TFBSs in cis-regulatory (promoter/enhancer/silencer) elements of the DNA are intensively studied to identify their effect on gene expression and thus their biological meaning. Motif discovery in biological sequences can be bioinformatically defined as the problem of finding short similar sequence elements shared by a set of nucleotides with a common biological function [11].

For *de novo* or known motif discovery within the previously identified peaks of your ChIP-Seq experiment *HOMER* provides the `findMotifsGenome.pl` program. This motif discovery algorithm uses "zero or one occurrence per sequence" (ZOOPS) scoring coupled with the hypergeometric enrichment calculations (or binomial) to determine motif enrichment comparing a peak set relative to another one. Many different output files will be placed in the defined output directory, including *html* pages showing the results.

findMotifsGenome.pl output files:

- homerMotifs.motifs<#>: Output files from the *de novo* motif finding, separated by motif length.

- homerMotifs.all.motifs: The concatenated file containing of all homerMotifs. motifs<#> files.
- motifFindingParameters.txt: Command used to execute findMotifsGenome.pl.
- knownResults.txt : Text file containing statistics about known motif enrichment.
- seq.autonorm.tsv: Autonormalization statistics.
- homerResults.html : Formatted output of *de novo* motif finding.
- homerResults/directory: Contains files for the homerResults.html webpage.
- knownResults.html: Formatted output of known motif finding.
- knownResults/directory: Contains files for the knownResults.html webpage.

Of course, there are countless other possibilities to further analyze ChIP-Seq data. To get a detailed description of all possible options of each *HOMER* software script just type the "command" of interest in the terminal (e.g., `mergePeaks`).

Take Home Message
- Epigenetic sequencing applications provide deep insights into the regulatory mechanisms of cells and tissue.
- ChIP-Seq can be performed to identify transcription factor binding sites, histone modifications, or DNA methylation, respectively.
- Different ChIP-Seq applications produce different type of peaks.
- Peak calling is often referred to the identification of enriched DNA regions compared to "Input" or "Control-IP" samples.
- *Sequencing Coverage and Depth* can be illustrated by the IGV or UCSC browser.
- The most important readouts of ChIP-Seq data analysis are: genomic feature association analysis, merge peak files with gene expression data (RNA-Seq), calculate ChIP-Seq Tag densities from different experiments, find motifs in peaks, and Gene Ontology Analysis.

Further Reading

- http://homer.ucsd.edu/homer/ngs/index.html
- https://www.bioconductor.org/help/course-materials/2016/CSAMA/lab-5-chipseq/Epigenetics.html
- Ma W, Wong WH. The analysis of ChIP-Seq data. Methods in enzymology. 2011.

Review Questions

Review Question 1

What can you learn by knowing the DNA binding sites of proteins such as transcription factors?

Review Question 2
   What is the primary purpose of chromatin sonication when performing a ChIP experiment?

Review Question 3
   Which of these is important for preparing templates for Next Generation Sequencing?

A. Isolating DNA from tissue.
B. Breaking DNA up into smaller fragments.
C. Checking the quality and quantity of the fragment library.
D. All of the above.

Answers to Review Questions

Answers to Question 1:
   A transcription factor recognizes and binds to specific sites in the genome, to recruit cofactors, and thus to regulate transcription. Thus ChIP allows identification of TF binding motifs and the direct downstream targets of a specific TF. Consequently, clustering of transcription-regulatory proteins at specific DNA sites can be assessed.

Answers to Question 2:
   Sonication of the chromatin is performed to reduce the size of the DNA fragments. Without this step, high molecular weight DNA would be immunoprecipitated, forming a large complex with the antibody. This would create loss of resolution and false positive results.

Answers to Question 3:
   D

Acknowledgements We thank Prof. Dr. Anja Bosserhoff (Institute of Biochemistry (Emil-Fischer Center), Friedrich-Alexander University Erlangen-Nürnberg) for reviewing this chapter and suggesting extremely relevant enhancements to the original manuscript.

References

1. Heinz S, Benner C, Spann N, Bertolino E, Lin YC, Laslo P, et al. Simple combinations of lineage-determining transcription factors prime cis-regulatory elements required for macrophage and B cell identities. Mol Cell. 2010;38(4):576–89.
2. Deliard S, Zhao J, Xia Q, Grant SF. Generation of high quality chromatin immunoprecipitation DNA template for high-throughput sequencing (ChIP-seq). J Vis Exp. 2013;(74):e50286.
3. Tian B, Yang J, Brasier AR. Two-step cross-linking for analysis of protein-chromatin interactions. Methods Mol Biol. 2012;809:105–20.
4. Jiang S, Mortazavi A. Integrating ChIP-seq with other functional genomics data. Brief Funct Genomics. 2018;17(2):104–15.

5. Pellegrini M, Ferrari R. Epigenetic analysis: ChIP-chip and ChIP-seq. Methods Mol Biol. 2012;802:377–87.
6. Langmead B, Trapnell C, Pop M, Salzberg SL. Ultrafast and memory-efficient alignment of short DNA sequences to the human genome. Genome Biol. 2009;10(3):R25.
7. Boeva V, Popova T, Bleakley K, Chiche P, Cappo J, Schleiermacher G, et al. Control-FREEC: a tool for assessing copy number and allelic content using next-generation sequencing data. Bioinformatics. 2012;28(3):423–5.
8. Quinlan AR. BEDTools: the Swiss-army tool for genome feature analysis. Curr Protoc Bioinformatics. 2014;47:11–2. 1–34
9. Quinlan AR, Hall IM. BEDTools: a flexible suite of utilities for comparing genomic features. Bioinformatics. 2010;26(6):841–2.
10. Kappelmann-Fenzl M, Gebhard C, Matthies AO, Kuphal S, Rehli M, Bosserhoff AK. C-Jun drives melanoma progression in PTEN wild type melanoma cells. Cell Death Dis. 2019;10 (8):584.
11. Zambelli F, Pesole G, Pavesi G. Motif discovery and transcription factor binding sites before and after the next-generation sequencing era. Brief Bioinform. 2013;14(2):225–37.

Appendix

Library Construction for NGS

Example Protocol: TruSeq Stranded Total RNA LT Sample Preparation Kit (Illumina)

1. Isolate total RNA
   - Determine quality and quantity of the isolated RNA.
   - RIN 9–10.

2. Preparations.
   - Use 1µg of Total RNA to initiate protocol (0.1–1.0µg starting range recommended).
   - Thaw frozen rRNA Binding Buffer [1], rRNA Removal Mix [2], rRNA Removal Beads & Elute Prime Fragment High Mix [7]. Place Resuspension Buffer at 4 °C for subsequent experiments.
   - Prepare 80% Ethanol [4] for AMPure XP bead washes.
   - Remove Elution Buffer [6], rRNA Removal Beads [3], and AMPure XP Beads [5] from 4 °C and bring to RT.
   - Rotate beads at 6 rpm.

3. Ribo-Zero Deplete and Fragment RNA (bind rRNA, rRNA removal, RNA clean-up, Depleted RNA Fragmentation)
   - Dilute 1ug of total RNA to 10µl using nuclease free H_2O in new 0.5 ml PCR tubes.
   - Add 5µl rRNA Binding Buffer to each sample.
   - Add 5µl of rRNA removal Mix.
   - Store rRNA Binding Buffer and rRNA Removal Mix at −20 °C.
   - Gently pipette the entire volume up and down to mix thoroughly.
   - Incubate in thermal cycler using the following profile to denature RNA:

© Springer Nature Switzerland AG 2021
M. Kappelmann-Fenzl (ed.), *Next Generation Sequencing and Data Analysis*, Learning
Materials in Biosciences, https://doi.org/10.1007/978-3-030-62490-3

| 68 °C | 5 min | Use heated lid at 100 °C |

- Remove samples from thermal cycler and incubate 1 min at RT.
- Vortex the warmed rRNA Removal Beads vigorously to completely resuspend the beads.
- Add 35μl of the rRNA Removal Beads to new 0.5 ml PCR tubes.
- Transfer entire contents of each sample (20μl) into the new tubes containing rRNA Removal Beads.
- Adjust pipette to 45μl, then with the tip of the pipette at the bottom of the tube, pipette quickly up and down 20 times to mix thoroughly.
- It is important to pipette up and down quickly to ensure thorough mixing. Insufficient mixing leads to lower levels of rRNA depletion.
- Prevent solution from foaming.
- Incubate the samples at RT for 1 min.
- Place the samples on the magnetic stand for 1 min. at RT.
- Transfer all supernatant from each sample into a new 0.5 ml PCR tube.
- Return rRNA Removal Beads [3] to 4 °C.
- Place the samples on the magnetic stand for 1 min at RT. (Repeat as necessary until there are no beads remaining.)
- Vortex the AMPure XP beads [5] until fully dispersed then add 99μl of AMPure XP beads [5] to each sample containing ribosomal depleted RNA. Pipette up and down gently 10× to mix.
- Incubate samples at RT for 15 min.
- Store AMPure XP Beads at 4 °C.
- Place the samples on the magnetic stand at RT for 5 min—make sure beads are completely deposited on side of tubes.
- Remove and discard all of the supernatant from each sample.
- Leave samples on the magnetic stand and wash wells with 200μl freshly prepared 80% ETOH [4]—DO NOT DISTURB BEADS.
- Incubate samples at RT°˜ 30 s. Remove ETOH using pipette—DO NOT DISTURB BEADS.
- Air dry samples on heatblock at 37 °C 2–5 min. (check beads in-between.)
- Briefly centrifuge thawed RT Elution Buffer [6] at 600×g for 5 s.
- Add 11μl Elution Buffer [6] to each sample and gently pipette entire volume 10×s to mix thoroughly.
- Store Elution Buffer at 4 °C.
- Incubate samples at RT for 2 min.
- Place samples on the magnetic stand for 5 min. at RT.
- Transfer 8.5μl of the supernatant into new 0.5 ml PCR tubes.

- Add 8.5µl Elute, Prime, Fragment High Mix [7] to each sample and gently pipette up and down 10× to mix thoroughly.
- Store Elute, Prime, Fragment High Mix at −20 °C.
- Incubate samples in the thermal cycler using the following profile:

94 °C	8 min	Use heated lid at 100 °C
4 °C		Hold

- Remove the samples when 4 °C is reached and centrifuge briefly.
- Proceed immediately to First Strand Synthesis.

4. Synthesize First Strand cDNA
   - Thaw one tube of First Strand Synthesis Act D Mix [8], spin briefly, and place on ice.
   - Add 1µl Superscript II [8] to 9µl First Strand Synthesis D Mix (8µl per sample needed; scale up volume as needed).
   - Mix well by finger flicking—DO NOT VORTEX.
   - Centrifuge briefly.
   - Store First Strand Synthesis Act D Mix at −20 °C immediately after use.
   - Add 8µl of Super Script II supplemented First Strand Synthesis Act D Mix to each sample.
   - Gently mix by pipetting up and down 6×.
   - Centrifuge briefly.
   - Incubate samples in a thermal cycler using the following profile:

25 °C	10 min	Use heated lid at 100 °C
42 °C	15 min	
70 °C	15 min	
4 °C	Hold	

- When the cycler reaches 4 °C remove samples and proceed immediately to the Second Strand Synthesis.

5. Synthesize Second Strand cDNA
   - Thaw Second Strand Marking Master Mix [9] at RT and warm 4 °C Resuspension Buffer to RT.
   - Warm AMPure XP Beads to RT° 30 min; rotate beads at 6 rpm.
   - Initiate thermal cycler: pre-heat to 16 °C.
   - Briefly Centrifuge Thawed Second Strand Marking Master Mix [9].
   - Add 5µl Resuspension buffer [10] into each sample.

- Add 20µl of thawed Second Strand Marking Master Mix [9] to each sample. Gently pipette up and down 6× to mix thoroughly.
- Store Second Strand Marking Master Mix at −20 °C.
- Incubate samples on a pre-heated thermal cycler with closed lid at 16 °C 1 h using the following profile:

| 16 °C | 1 h | *Do not* use heated lid (program: HEADER) |
| 16 °C | Hold | |

- Remove samples and place at RT to equilibrate.
- Vortex the RT AMPure XP beads until fully dispersed then add 90µl of beads to each sample containing 50µl of ds cDNA. Pipette up and down gently 10×s to mix.
- Incubate samples at RT 15 min.
- Place samples on the magnetic stand at RT for 5 min—make sure beads are completely deposited on side of tubes.
- Remove and discard 135µl of the supernatant from each sample.
- Some liquid may remain in the tubes—DO NOT DISTURB BEADS.
- Leave samples on the magnetic stand and wash with 200µl freshly prepared 80% EtOH—DO NOT DISTURB BEADS.
- Incubate samples at RT 30 s. Remove ETOH using pipette—DO NOT DISTURB BEADS.
- Repeat 80% ETOH wash.
- Air dry samples on heatblock at 37 °C 2–5 min. (check beads in-between.)
- Add 17.5µl Resuspension Buffer [10] to each sample. Mix up and down 10× to completely resuspend beads.
- Incubate the samples at RT for 2 min.
- Place samples on the Magnetic Stand at RT°~ for 5 min.
- Transfer 15µl of the supernatant containing the ds cDNA to new 0.5 ml tubes.
- Some liquid may remain in the wells—DO NOT DISTURB BEADS.

Time until this point: ~7 h
Samples can be stored at this point at −20 °C for up to 7 days.

6. Adenylate 3′ Ends
   - Remove one tube of A-Tailing Mix [11] from −20 °C and thaw at RT.
   - A-Tailing Mix aliquots (á 50µl) and label tubes. Store at −20 °C.
   - Remove samples from −20 °C storage and thaw at RT. Briefly centrifuge samples at 280×g for 1 min.
   - Initiate thermal cycler profile: pre-heat to 37 °C.
   - Add 2.5µl Resuspension Buffer [10] to each sample.

- Add 12.5µl of thawed A-Tailing Mix [11] to each sample. Pipette up and down 10×
  to mix.
- Incubate samples in the pre-heated thermal cycler with closed lid at 37 °C for 30 min
  using the following profile:

37 °C	30 min	Use heated lid at 100 °C
70 °C	5 min	
4 °C	Hold	

- Immediately remove the samples from the thermal cycler—Immediately proceed with
  Adapter Ligation!

7. Ligate Adapters
   - Remove the appropriate RNA Adapter Index tubes (AR001-AR012, depending on
     the RNA Adapter Indexes being used) and one tube of Stop Ligase Buffer
     [13] from −20 °C and thaw at RT.
   - Remove Resuspension Buffer from 4 °C and warm to RT.
   - Remove the AMPure XP Beads from 4 °C storage and warm to RT at least 30 min.
   - Initiate thermal cycler profile: pre-heat to 30 °C.
   - Prepare new 0.5 ml PCR tubes (2×, for Adapter ligation and PCR).
   - Briefly centrifuge the thawed RNA Adapter Index tubes and Stop Ligase Mix.
   - Add 2.5µl Resuspension Buffer [10] to each sample.
   - Remove the DNA Ligase Mix [12] from −20 °C immediately before use and leave in
     −20 °C benchtop storage cooler.
   - Add 2.5µl of DNA Ligase Mix [12] directly from the −20 °C benchtop storage cooler
     to each sample.
   - Return −20 ° C benchtop storage cooler back to freezer immediately after use.
   - Add 2.5µl of the appropriate thawed RNA Adapter Index (AR001-AR012) to each
     sample.
   - Adjust the pipette to 40µl and gently pipette the entire volume up and down 10× to
     mix thoroughly.
   - Change gloves after pipetting each adapter and clean pipette.
   - Immediately store adapters at −20 °C (sign as used).
   - Incubate samples in the pre-heated thermal cycler with closed lid at 30 °C for 10 min
     using the following profile:

30 °C	10 min	Use heated lid at 100 °C
30 °C	Hold	

- Remove samples from the thermal cycler.
- Add 5μl Stop Ligase Buffer [13] to each sample and mix by gently pipetting up and down 10×.
- Vortex the pre-warmed AMPure XP Beads until fully dispersed.
- Add 42μl of beads to each sample. Gently pipette up and down 10× to mix well.
- Incubate samples at RT 15 min.
- Place samples on the magnetic stand at RT for at least 5 min making sure liquid clears.
- Remove and discard 79.5μl of supernatant from each sample. DO NOT DISTURB BEADS. Change tips after each removal.
- Leave samples on the magnetic stand while performing the 80% ETOH washing steps.
- Add 200μl freshly prepared 80% EtOH to each sample without disturbing the beads.
- Incubate samples at RT for at least 30 s. Then remove and discard all the supernatant from each sample without disturbing the beads. Change tips between wells.
- Repeat ETOH wash.
- Air dry samples on heatblock at 37 °C 2–5 min. (check beads in-between.)
- Remove samples from magnetic stand and resuspend the dry pellet in each sample with 52.5μl of Resuspension Buffer. Gently pipette up and down 10×s to mix.
- Incubate samples at RT for 2 min.
- Place samples on the magnetic stand and incubate at least 5 min—make sure liquid clears.
- Transfer 50μl of clear supernatant from each sample to the corresponding new 0.5 ml tubes. Some residual liquid may remain in each sample.
- Vortex the AMPure XP beads to disperse and add 50μl of beads into each sample for a second clean-up. Pipette up and down 10× to mix thoroughly.
- Incubate samples at RT for 15 min.
- Place the samples on the magnetic stand at RT for 5 min making sure the liquid has cleared.
- Remove and discard 95μl of the supernatant from each sample without disturbing the beads. Some liquid may remain in each sample. Remember to change tips between wells.
- Leave samples on the magnetic stand while performing the 80% ETOH washing steps.
- Add 200μl freshly prepared 80% EtOH to each sample without disturbing the beads.
- Incubate samples at RT for at least 30 s, then remove and discard all of the supernatant from each sample without disturbing beads. Change tips after each removal.
- Repeat ETOH wash.
- Air dry samples on heatblock at 37 °C 2–5 min. (check beads in-between.)
- Remove samples from magnetic stand and resuspend the dry pellet in each sample with 22.5μl of Resuspension Buffer. Gently pipette up and down 10× to mix.
- Incubate samples at RT 2 min.
- Place samples on the magnetic stand at RT for at least 5 min making sure liquid clears.
- Transfer 20μl of clear supernatant from each sample to the corresponding new 0.5 ml PCR tubes (PCR). Some residual liquid may remain in each sample.

Samples can be stored at this point at −20 °C for up to 7 days.

8. Enrich DNA Fragments.
   - Remove one tube each of PCR Master Mix and PCR Primer Cocktail from −20 °C and thaw at RT—centrifuge briefly.
   - Remove the AMPure XP Beads from 4 °C and warm to RT at least 30 min.
   - Remove Resuspension Buffer and samples from −20 °C and thaw to RT. Briefly centrifuge samples at 280×g for 1 min.
   - Prepare new 0.5 ml PCR tubes.
   - Aliquot the appropriate volume of each reagent (with 10% excess per tube) into strip tubes. Cap tubes and keep on ice until needed. Remaining content can be re-store at −20 °C.
   - Add 5μl of thawed PCR Primer Cocktail to each sample.
   - Add 25μl of thawed PCR Master Mix to each sample. Pipette up and down 10× to mix thoroughly.
   - Run PCR in the pre-heated thermal cycler with closed lid using the following profile:

1 cycle		
98 °C	30 s	Use heated lid at 100 °C
15 cycles		
98 °C	10 s	
60 °C	30 s	
72 °C	30 s	
1 cycle		
72 °C	5 min	
4 °C	Hold	

   - Vortex the pre-warmed AMPure Beads until they are completely dispersed.
   - Add 50μl of the beads to each sample containing 50μl of PCR amplified library and pipette up and down 10× to mix thoroughly.
   - Incubate the PCR tubes at RT 15 min.
   - Transfer samples to the magnetic stand at RT for at least 5 min making sure the liquid clears.
   - Remove and discard 95μl of the supernatant from each sample without disturbing the beads. Some liquid may remain in the wells. Remember to change tips.
   - Leave samples on the magnetic stand while performing the 80% ETOH washing steps.
   - Add 200μl of freshly prepared 80% EtOH to each sample without disturbing the beads.
   - Incubate the samples at RT at least 30 s, then remove and discard all the supernatant from each sample without disturbing the beads. Remember to change tips.
   - Repeat ETOH wash.

- Incubate the samples open on the magnetic stand at RT 15 min to dry residual ETOH.
- Remove samples from magnetic stand and resuspend the dry pellet in each sample with 32.5μl of Resuspension Buffer. Gently pipette up and down 10× to mix.
- Incubate samples at RT for 2 min.
- Place the samples on the magnetic stand at RT for at least 5 min making sure liquid clears.
- Transfer 30μl of clear supernatant from each sample to the corresponding 1.5 mL non-sticky Sarstedt microcentrifuge tube. Some residual liquid may remain in each sample.

Amplified libraries can be stored at −20 °C for up to 7 days

9. Validate Library
   - Determine the concentration of each amplified library using the Nanodrop.
   - Perform QC of the amplified library by running 1μl of each sample on the Agilent 4200 Bioanalyzer using the Agilent DNA 1000 Chip.
   - The final product should be a band at approximately 280 bp for a single read library.
   - Calculate the nM concentration of each library.

Example Protocol: TruSeq ChIP Library Prep Kit (Illumina)

- Verify the size distribution of each ChIP DNA sample by running a 2μl aliquot on Agilent High Sensitivity DNA chip using an Agilent Technologies TapeStation.
- Quantify 1μl of each ChIP DNA sample using a Qubit 1× dsDNA HS Assay Kit (invitrogen: Q33231).
- Illumina recommends normalizing the ChIP DNA samples to a final volume of 50μl at 100–200 pg/μl.
- Remove the AMPure XP beads from storage and let stand for at least 30 min to bring them to room temperature.
- Pre-heat the thermal cycler to 30 °C.
- Choose the thermal cycler pre-heat lid option and set to 100 °C.
- Freshly prepare 70% ethanol (with sterile Distilled Water (DNase/Rnase Free).
- Remove the A-Tailing Mix from −15 °C to −25 °C storage and thaw at room temperature.

1. Polish ChIP DNA ends
   - Use 50μl ChIP sample.
   - For Input samples, dilute 10 ng into 50μl total volume (ddH₂O).

Master mix for blunting DNA

	1 x [µl]	x [µl]
DNA		
Resuspension Buffer [1]	10	
End Repair Mix [2]	40	
	100.00	

- 50µl sample
- +50µl master mix
- 100µl   → 30 min @ 30 °C in thermocycler

2. Clean up with AMPure beads
   - Vortex the AMPure XP Beads [5] until they are fully dispersed.
   - 100µl after polish DNA ends.
   - +160µl AMPure XP magnetic beads.
   - Gently pipet up and down 10×.
   - 5 min at room temperature.
   - Place tubes on a magnet holder for 2 min (or until the liquid is clear).
   - Discard 127.5µl supernatant.
   - Again discard 127.5µl supernatant (not once 255µl because of the suck).
   - 2 × 500µl of 70% ethanol for 30 s, wash magnetic beads, discard supernatant.
   - Air dry the beads on magnetic stand at RT C for 2–5 min or at 37 °C in a heatblock.
   - Elute with 17.5µl Resuspension Buffer, pipette the entire volume up and down 10 times to mix thoroughly.
   - 2 min at room temperature.
   - ⎗ 13,000 rpm, 1 s.
   - Place tubes on a magnet holder for 2 min.
   - Transfer 15µl in new tube.

3. Perform 3′-dA addition

Master mix for A-tailing DNA

	1 x [µl]	x [µl]
Resuspension Buffer[1]	2.50	
A-Tailing Mix[3]	12,5	
	15	

15µl sample
+ *15µl* master mix
30µl

PCR-cycler:

Pre-heated lid 00 °C	37 °C	30 min
	70 °C	5 min
	4 °C	∞

4. Adapter Ligation
   - Preparation of adapters (RNA Adapter Indices: AR001–AR016, AR018–AR023, AR025, AR027):
   - (at the moment Adapters of the TruSeq RNA-Seq-Kit).
   - Centrifuge the Stop Ligation Buffer and appropriate/desired thawed RNA Adapter tubes to 600 ×g for 5 s.
   - Immediately before use, remove the Ligation Mix tube from −20 °C and to −20 °C immediately after use.
   - 30µl sample.
   - +2.5µl Resuspension Buffer[1].
   - +2.5µl Ligation Mix[4].
   - +2.5µl thawed RNA Adapter Index.
   - 37.5µl.
   - Gently pipette the entire volume up and down 10× to mix thoroughly.
   - Centrifuge PCR samples at 280 g for 1 min.
   - → ligate for 10 min at 30 °C (thermal cycler use heated lid at 100 °C?)
   - +5µl Stop Ligation Buffer [6]
   - Gently pipette the entire volume (42,5µl) up and down 10 times.
     - +7.5µl H$_2$O
     - 50µl.
   - Gently pipette the entire volume (50µl) up and down 10 times.

5. Clean up with AMPure beads.
   - 50µl after ligation.
   - +55µl AMPure XP magnetic beads.
   - Gently pipet up and down 10 x.
   - 5 min at room temperature.
   - Place tubes on a magnet holder for 2 min (or until the liquid is clear).
   - Discard supernatant.
   - 2 × 500µl of 70% ethanol for 30 s, wash magnetic beads, discard supernatant.
   - Air dry the beads on magnetic stand at RT for 2–5 min or at 37 °C in a heatblock.
   - Elute with 50 µl H$_2$O, pipette the entire volume up and down 10 times to mix thoroughly.
   - 2 min at room temperature.
   - ⚙ 13,000 rpm, 1 s.

- Place tubes on a magnet holder for 2 min (or until the liquid is clear).
- Transfer 50µl in new tube for second clean-up.

6. Clean up with AMPure beads
   - 50µl after ligation.
   - + 55µl AMPure XP magnetic beads.
   - Gently pipet up and down 10×.
   - 5 min at room temperature.
   - Place tubes on a magnet holder for 2 min (or until the liquid is clear).
   - Discard supernatant.
   - 2 × 500µl of 70% ethanol for 30 s, wash magnetic beads, discard supernatant.
   - Air dry the beads on magnetic stand at RT for 2–5 min or at 37 °C in a heatblock.
   - Elute with 20 µl H2O, pipette the entire volume up and down 10 times to mix thoroughly.
   - 2 min at room temperature.
   - ⊠ 13,000 rpm, 1 s.
   - Place tubes on a magnet holder for 2 min (or until the liquid is clear).
   - Transfer 20µl in new tube for PCR (0.5 ml PCR tube).

7. Enrich DNA Fragments

Master mix final PCR (ChIP)

	1 x [µl]	x [µl]
PCR Primer Cocktail[7]	5.00	
PCR Primer Master Mix[8]	25.00	
	30.00	

20µl DNA sample
+*30µl* master mix
50µl.

PCR-cycler:

	98 °C	30 s
4×	98 °C	10 s
	60 °C	30 s
	72 °C	30 s
	72 °C	5 min
	4 °C	∞

8. Clean up with AMPure beads.
   - 50µl after PCR.
   - + 90µl AMPure XP magnetic beads.
   - Gently pipet up and down 10×.
   - 5 min at room temperature.
   - Place tubes on a magnet holder for 2 min (or until the liquid is clear).
   - Discard supernatant.
   - 2 × 500µl of 70% ethanol for 30 s, wash magnetic beads, discard supernatant.
   - Air dry the beads on magnetic stand at RT for 2–5 min or at 37 °C in a heatblock.
   - Elute with 30µl H$_2$O, pipette the entire volume up and down 10 times to mix thoroughly.
   - 2 min at room temperature.
   - ⬛ 13,000 rpm, 1 s.
   - Place tubes on a magnet holder for 2 min (or until the liquid is clear).
   - Transfer 30µl in new tube for Blue Pippin.

9. BLUE PIPPIN (size selection of ChIP Seq Library)

Purify and Size Select the Ligation Products
- Perform the size selection using the Sage BLUE PIPPIN.
- Bring Pippin Prep loading solution to RT.

Prepare DNA Samples for loading
- Initiate the BluePippin instrumentation allowing software to launch.
- Add 10µl of RT loading solution to each sample for a total of 40µl. Mix well and centrifuge briefly.

Program Protocol
- From main screen select "Protocol" to go to the Protocol Editor Screen. Press "New" to open a new protocol.
- Select the appropriate cassette type.
- Select a reference lane which will contain the DNA marker.
- NOTE: Verify lane assignment of software display vs cassette lanes.
- Enter - bp target size 275 bp.
- -bp start size 250 bp. NOTE: Cut size may need to be adjusted to target optimal size distribution of fragmented sample.
- -bp end size 300 bp.
- Enter sample name into the "Sample ID" field.
- Select "Save as" and name protocol to retain for future applications (TruSeq ChIP 250–300 bp).
- Return to Main Screen.

Prepare Cassette
- Remove cassette from packaging and wipe off excess moisture. Verify correct cassette type is used.
- Visually inspect buffer chamber and gel columns. From side view ensure buffer chamber volume is more than ½ full. Inspect gel column for breakage, bubbles, or gaps. Any lanes with these defects must be abandoned. Remaining lanes are still functional.
- Inspect for air gaps in and behind elution wells by tilting cassette to the left (loading well end). Tap on bench lightly to dislodge.
- Install cassette into tray of Pippin Prep with loading wells to the left.
- Remove sealing tape without tipping the cassette.
- Remove all buffer from elution well and replace with 40µl of electrophoresis buffer.
- Test electrophoretic current continuity:
- Close sliding lid on instrument.
- In the run screen select "Manual Mode" from the "Protocol Name" drop down menu.
- Press "Test" in controller on the Main Screen. Test takes 30 s. Finishing with "Pass" of "Fail" message.
- If test fails—in the separation lane: lane unusable—no recovery.
- In the elution lane: Verify 40µl of running buffer is present in the elution well and re-test. If it fails again, lane can be used for reference sample.
- Cover all collection wells with tape to minimize recover volume to ~50µl. Using razor blade cut tape between elution wells so that.
- individual tape segments can be removed to minimize potential cross contamination.

Sample Loading
- Verify that sample wells are completely full. Top off wells with electrophoresis buffer if necessary. Total well volume is 70µl.
- Remove 40µl of running buffer from the loading well leaving ~30µl in well.
- Load 40µl of pre-mixed Pippin Prep DNA Marker into the sample lane that has been assigned for the reference. (Reference lane can be changed as needed.)
- Load 40µl of the supplemented sample into the appropriate lane loading well.
- Close lid.
- Select desired protocol from the "Protocol Name" drop down menu. (i.e.: TruSeq ChIP 250–300 bp.)
- Press "Start."

Sample Collection
- When complete, remove elution well tape and transfer all samples from sample elution well to a labeled microcentrifuge tube.
- Determine volume of collected sample. Volume can be from 40–70µl.

- Clean electrodes of Pippin Prep by installing cleaning cassette filled with H2O and closing lid for 30 s. Remove Cassette.

10. Sample Clean-Up
   - Use the QiagenQIAquick PCR Purification Kit (cat# 28106) for clean-up procedure. Perform the PCR Purification Protocol.
   - Add 5 volumes of PB buffer (with Indicator) to 1 volume of the Pippin Prep elute (250μl PB + I buffer +50μl Pippin Prep Elute).
   - Check the color of the mixture and ensure that it is yellow (similar to the original color of buffer ERC).
   - If color is orange or violet, add 10μl of 3 M sodium acetate pH 5.0 and mix. The color will return to yellow.
   - Place a MinElute column in a 2 ml collection tube.
   - Bind DNA by adding the sample to the MinElute column and centrifuging 1 min at 13,000 RPM at RT°~.
   - Discard the flow-through and place the MinElute column back in the same collection tube.
   - Wash column by adding 750μl of Buffer PE to the MinElute column and centrifuge 1 min at 13,000 RPM at RT°~.
   - Discard flow-through and place MinElute column back in the same collection tube.
   - Centrifuge MinElute column for an additional 1 min to remove any residual wash buffer PE.
   - Place the MinElute Column in a new labeled 1.5 ml microcentrifuge tube.
   - Elute DNA by adding 23μl of EB buffer (10 mM Tris-HCl pH 8.5) to the center of the membrane. Let column stands at RT°~ 1 min.
   - Centrifuge MinElute column for 1 min at 13,000 RPM.
   - Check volume of eluted samples and verify that 20μl + are obtained. Add resuspension buffer, if needed, to obtain 20μl.

11. Enrich DNA Fragments.

Master mix final PCR (ChIP)		
	1 x [μl]	x [μl]
PCR Primer Cocktail[7]	5.00	
PCR Primer Master Mix[8]	25.00	
	30.00	

20μl sized DNA sample
+ *30μl* master mix
50μl.

PCR-cycler:

	98 °C	30 s
12×	98 °C	10 s
	60 °C	30 s
	72 °C	30 s
	72 °C	5 min
	4 °C	∞

12. Clean up with AMPure beads
    • 50μl after ligation.
    • +55μl AMPure XP magnetic beads.
    • Gently pipet up and down 10×.
    • 5 min at room temperature.
    • Place tubes on a magnet holder for 2 min (or until the liquid is clear).
    • Discard supernatant.
    • 2 × 500μl of 70% ethanol for 30 s, wash magnetic beads, discard supernatant.
    • Air dry the beads on magnetic stand at RT for 2–5 min or at 37 °C in a heatblock.
    • Elute with 50 μl H$_2$O, pipette the entire volume up and down 10 times to mix thoroughly.
    • 2 min at room temperature.
    • ⏲ 13,000 rpm, 1 s.
    • Place tubes on a magnet holder for 2 min (or until the liquid is clear).
    • Transfer 50μl in new tube for second clean-up.
13. Clean up with AMPure beads.
    • 50μl after ligation.
    • +55μl AMPure XP magnetic beads.
    • Gently pipet up and down 10×.
    • 5 min at room temperature.
    • Place tubes on a magnet holder for 2 min (or until the liquid is clear).
    • Discard supernatant.
        2 × 500μl of 70% ethanol for 30 s, wash magnetic beads, discard supernatant.
    • Air dry the beads on magnetic stand at RT for 2–5 min or at 37 °C in a heatblock.
    • Elute with 16μl EB, pipette the entire volume up and down 10 times to mix thoroughly.
    • 2 min at room temperature.
    • ⏲ 13,000 rpm, 1 s.

- Place tubes on a magnet holder for 2 min (or until the liquid is clear).
- Transfer 16µl in new tube.

14. TapeStation (Quality and quantity control)

15. Pool libraries (Qubit).
    - *store libraries @ −20 °C.*

NGS Technologies

Table 1 Major NGS platforms and their general properties

Platform	Company	Instrument	Reads per run	Average read length [bp]	Read type	Error RATE	Data generated per run (Gb)	Template (single molecule or clonal)	Sequencing scheme	Cyclic or continous	Vizualization	Sequencing principle
Sanger sequencing	ThermoFisher/ Roche	ABI 3500/ 3730	//	Up to 1 kb	Primer defined	//	0,0003	Clonal	//	//	Optical, electrophoresis	//
		MiniSeq	25 Mio.	1 × 75 to 2 × 150	Short	1/1000	1.7–7.5	Clonal	Surface or microwells	Cyclic	Optical	Terminated synthesis
		MiSeq	25 Mio.	1 × 36 to 2 × 300	Short	1/1000	0.3–15	Clonal	Surface or microwells	Cyclic	Optical	Terminated synthesis
		NextSeq 2000	400 Mio.	1 × 75 to 2 × 150	Short	1/1000	10–120	Clonal	Surface or microwells	Cyclic	Optical	Terminated synthesis
		HiSeq 4000	5 bio.	1 × 50 to 2 × 250	Short	1/1000	10–1000	Clonal	Surface or microwells	Cyclic	Optical	Terminated synthesis
		NovaSeq 5000/6000	20 bio.	2 × 50 to 2 × 150	Short	1/1000	2000–6000	Clonal	Surface or microwells	Cyclic	Optical	Terminated synthesis
IonTorrent	ThermoFischer	PGM	5.5 Mio.	Up to 400	Short	~2.5%	0.08–2	Clonal	Beads in wells	Cyclic	Electrical	Unterminated synthesis
		S5	2–130 Mio.	Up to 400	Short	~1%	0.6–15	Clonal	Beads in wells	Cyclic	Electrical	Unterminated synthesis
		Proton	40–80 Mio.	Up to 200	Short	~1%	10–15	Clonal	Beads in wells	Cyclic	Electrical	Unterminated synthesis
Pacific Bioscience	Pacific Bioscience	PacBio RSII	50 T	Up to 60 kb	Long	1/100	10	Single molecule	Microwells	Continous	Optical	Unterminated synthesis
	Illumina	Sequel	500 T	Up to 60 kb	Long	1/100	15	Single molecule	Microwells	Continous	Optical	Unterminated synthesis
		Sequel II	4 Mio.	Up to 60 kb	Long	1/100	100	Single molecule	Microwells	Continous	Optical	Unterminated synthesis

(continued)

Table 1 (continued)

Platform	Company	Instrument	Reads per run	Average read length [bp]	Read type	Error RATE	Data generated per run (Gb)	Template (single molecule or clonal)	Sequencing scheme	Cyclic or continous	Vizualization	Sequencing principle
Oxford Nanopore	Oxford Nanopore	MInION	7–12 Mio.	Up to 100 kb	Long	5–15%	50	Single molecule	Membrane	Continous	Electrical	Pore passing
		PromethION	160–550 Mio.	Up to 100 kb	Long	5–15%	220	Single molecule	Membrane	Continous	Electrical	Pore passing
		Flongle	3 Mio.	Up to 100 kb	Long	5–15%	2	Single molecule	Membrane	Continous	Electrical	Pore passing
		GridION	35–60 Mio.	Up to 100 kb	Long	5–15%	50	Single molecule	Membrane	Continous	Electrical	Pore passing
10× Genomics	10× Genomics	Chromium, illumina	See illumina									

Index

© Springer Nature Switzerland AG 2021
M. Kappelmann-Fenzl (ed.), *Next Generation Sequencing and Data Analysis*, Learning
Materials in Biosciences, https://doi.org/10.1007/978-3-030-62490-3

Printed in the United States
by Baker & Taylor Publisher Services